SpringerBriefs in Aging

More information about this series at http://www.springer.com/series/10048

Sherry Cummings · Nancy P. Kropf

Senior Cohousing

A New Way Forward for Active Older Adults

Sherry Cummings
College of Social Work
University of Tennessee
Knoxville, TN, USA

Nancy P. Kropf
School of Social Work, Perimeter College
Georgia State University
Atlanta, GA, USA

ISSN 2211-3231 ISSN 2211-324X (electronic)
SpringerBriefs in Aging
ISBN 978-3-030-25361-5 ISBN 978-3-030-25362-2 (eBook)
https://doi.org/10.1007/978-3-030-25362-2

This Springer imprint is published by the registered company Springer Nature Switzerland AG
The registered company address is: Gewerbestrasse 11, 6330 Cham, Switzerland

During our travels to twelve senior cohousing communities, the two authors spoke with 76 of residents. These women and men opened their homes, shared their experiences, and made this project informative …… and fun!

We dedicate this book to those individuals as a small way to thank them for their kindness, and generosity of spirit.

Contents

Abbreviations

EFF	Elder Family Fellowship
NOBO	North Boulder
SCCs	Senior cohousing communities

Chapter 1
Introduction

> …. we're on the leading edge of the baby boomers so we don't do anything like anybody has ever done before and that includes aging. You know, we've seen our parents in nursing homes and that's not where we want to go…. (Tammy)

Our society is currently experiencing demographic shifts due to the aging of the baby boom generation, and Tammy (above) captures well the magnitude of change brought about by the baby boomers. This cohort, born after World War II during the years of 1946–64, has transformed social institutions and trends at every point in the life course. As members of this cohort ourselves, both of the authors have faced the changes firsthand—and many of you who are reading this book have gone through these experiences as well. We overflowed school classrooms as our sheer number went beyond the capacity of space and resources. We embraced new trends and modified social values and norms in many areas including choices, customs, and social roles. Now as we grow older, we are transforming options for later life by our preferences and decisions.

We see these trends in many ways which create new and emerging journeys into later life. One is the way that we care for ourselves. For example, greater attention is being paid to staying healthy and fit into later life such as described in the book *Fitness After Fifty: Eat Well, Move Well, Be Well* (Rosenbloom & Murray, 2018). Take the experience of running a marathon, for example. Previously, this event was dominated by younger runners who were viewed as having the endurance to complete the 26.2 miles. However, there has been a shift in runners' profiles as greater numbers of master runners compete in this sport. In fact, now over 50% of men and 40% of women are over the age of 40, and often runners are competing into their 60s or 70s with substantial benefits to physical health (McMahan, 2015).

S. Cummings and N. P. Kropf, *Senior Cohousing*, SpringerBriefs in Aging,
https://doi.org/10.1007/978-3-030-25362-2_1

Even with the onset of chronic conditions, people are also looking more holistically at their health and well-being. In older adults with diabetes, for example, one report indicated that 25% had used complimentary alternative medicine treatments within the past year (Rhee, Westberg, & Harris, 2018). These included interventions such as acupuncture, massage, herbs and supplements, and mind-body programs (e.g., yoga and tai-chi). Spiritual changes are also happening. The baby boom generation embraces a more diverse array of spiritual traditions, beyond the mainstream religions, which includes meditation, connecting with nature, and other practices.

Another area that is changing is the decision about where and how to live during later years. Many older adults state a preference for living in their own homes; however, this decision can have unintended consequences such as worrying about how to manage if a health crisis is experienced, or feeling unsafe in a home or neighborhood. Living situations that provide people with companionship, ways to stay engaged with others, as well as provide a purpose and community have emerged. As baby boomers consider later life, many are exploring options that channel relationship and support. In this way, the baby boom generation is once again transforming options for later life—just as we have done during early times in our lives!

The desire for connection is an important factor in exploring more communal living arrangements. Isolation is particularly acute in later adulthood when there are fewer options for increasing engagement and social ties. Limited mobility, sensory declines, and other age-related changes may prevent older adults from establishing meaningful and supportive relationships. Additionally, later life is without some of the settings that provide a context for establishing friendships such as employment and school settings. These combined factors contribute to loneliness, depression, and disengagement. Consider a story in the New York Times (Onishi, 2017) about isolation in older adults in Japan, which is one of the top countries for long life expectancies. Mega-apartment complexes house scores of older adults, each who live alone within her or his small unit. Without meaningful interaction within these buildings, individuals stay cocooned within their own units- never knowing their neighbors. A shocking tale brought this situation to national attention when a 69 year old man was found dead in his apartment, *being deceased for three years* prior to being discovered. This is an extreme example, yet it does speak to both a concern shared by many about potential isolation and the desire to be involved in engaged and caring relationships in later life

1.1 Senior Cohousing

In response to desiring connection and engagement, senior cohousing communities (SCCs) have been developed to bring together individuals around shared values[1]. While cohousing was initially designed as multigenerational communities, a move-

[1]For more information about Intentional Communities, visit the Fellowship for Intentional Community for references and resources: https://www.ic.org/.

Fig. 1.1 Elder Family Fellowship

ment has formed to establish communities specifically for those in later life (Fig. 1.1). A more detailed history of cohousing is provided in Chap. 2.

Senior cohousing communities offer living arrangements which provide space and attention to the experience of growing older, including both the rewarding aspects and the challenges. In multigenerational cohousing, these issue are diluted by many important ones that are part of family life such as raising children and labor force issues. One of the women interviewed as part of this project summed up her decision to move to a SCC instead of a multigenerational community:

> I had good friends that lived there [multigenerational cohousing community], and I observed the interaction of the community, intergenerational, and it seemed that there were issues involving young children, and different things that older people would not necessarily be concerned about. That's why I then thought that I would concentrate on a senior community, because we wouldn't have some of those issues.

In addition to having space for issues of aging, SCCs promote a sense of facing the aging process collectively. In these communities, there is a perception that everyone is sharing the aging process together which brings security and a sense of caring (Fig. 1.2). Noreen summed up her impression of living in an SCC with this view, "I think that there's a sense of more ease with it knowing that we're not gonna be alone. Knowing that there are people around who care. So, I think for me, that makes me feel just more confident about the [aging] process. That's the way I wanna live and the way I wanna die."

Like cohousing communities in general, SCCs are founded upon shared values and assumptions that provide the basis for residents. The six principles of cohousing are:

1. **Participatory Process**—development of a community is started with a vision statement. Future residents lead the development process and are co-collaborators in overall design of the community.
2. **Neighborhood Design**—the lay out and configuration of the space within the community is created to stress connection and sense of community
3. **Common Facilities**—a common building on the property includes kitchen space for shared meals, along with other spaces, such as laundry and library, based upon the vision and shared principles
4. **Resident Management**—decisions about division of labor are made by the community residents themselves. While residents have responsibility for their own units, decisions about governing policies and work related to common spaces (common building, shared meals, gardens, patios) are determined by the residents
5. **Non-hierarchical Structure and Decision-making**—most cohousing communities use a consensus decision making model. There is no person or group that has greater power or decision making over others
6. **No Shared Community Economy**—the community does not exist to create an income stream for the residents.

Fig. 1.2 Socializing in the common area

1.2 The Purpose of the Book

This book was written to specifically explore older adults' experience of living in a senior cohousing community. The idea came to the authors through both professional and personal perspectives. Professionally, both of us are social workers and gerontologists and we have studied numerous issues of later life including mental health, caregiving, and social networks. As teachers, we have also worked with students to learn effective practice with the older population—and this includes eradicating some of the myths of growing older. One common myth is that all older adults end up in nursing homes. In exploring SCCs, we hope to bring additional understanding to social work, gerontology and other students who seek careers in health and human service about how these communities enhance options for adults as they age.

Along with this professional journey, however, there are personal reasons for our interest. As baby boomers, we are exploring options for our own aging years. What kind of place *would we* like to live? Where do *we see ourselves* aging? These are important questions and are informed by caregiving experiences that we have had with our own parents. We both have had a parent who lived in a long-term care setting and concluded that this would not be a place that we would choose to move. As true academics, we decided to address this issue by exploring other living options—and spent our summer researching, visiting, and interviewing residents of SCCs across the country.

During our visits, we were interested in several questions. How did the residents make the important decision to move into this type of residential setting? What activities do they participate in—both within the SCC and in the surrounding environment? As they look into the future, how do they see their aging within this type of community? What are the biggest challenges, and what has this experience taught them about themselves? We addressed the same set of questions across all of the communities to compare similar and unique issues that arose. (A full set of the questions can be found in Appendix A).

We selected twelve SCCs to visit. We used a sampling approach called maximum variation purposive sampling to identify 12 SCCs that would represent the full variety of senior cohousing characteristics. The purposive sampling was based on the variation of SCCS along the following dimensions: (1) Mission; (2) Population; (3) Structure; and (4) Geographic location. The communities identified were distributed across California, Colorado, New Mexico, North Carolina, Virginia and Washington DC. In some communities, we interviewed residents separately and in other places within group formats—whatever made the residents most comfortable. All in all, we ended up interviewing a total of 76 residents. In a few places, we were invited to spend the night and participate in some of the activities such as meals, happy hour, vespers and soaking in a hot tub. In all cases, the residents with whom we met were generous with their time, and honest with their answers.

This book tells the story of life in senior cohousing communities. We employed an existential-phenomenological qualitative research method. The ultimate goal in using this approach is to gain knowledge of the phenomena being studied by achieving a deeper understanding of the lived experience of the individual or group of individuals involved (Collingridge & Gantt, 2008). Informed consent was obtained for each person prior to being interviewed. Participants were told that no real names would be used (all names appearing in this book are aliases) and that no specific remarks would be attributed to those living in particular communities. In this way, we hoped to preserve the anonymity of those interviewed and encourage the most open and honest answers to the questions posed. Data for this study consisted of the transcripts made from the audio-recorded focus group and individual interviews. The transcripts were then read individually by each of the authors and analyzed to discern common themes that emerged in response to the questions asked. Final themes were discussed and agreed on by the researchers. What is presented in this book, therefore, is not an account of individual senior cohousing communities. Rather, it is the story of common experiences of those who live in SCCs regardless of the geographic area in which the community is located, its size, or years in existence. Our experience highlights some of the common characteristics found across the settings which include a desire to be in relation to others, the offering and receiving of assistance and support, and a common structure that provides for ongoing maintenance of communal space. As you read the stories of the residents and the SCCs, you will find powerful descriptions of resilience, connection, and caring. One of the men interviewed was in the process of moving to a particular community, and summed up the benefits as he made this decision:

> … being here, knowing that there is an intact support system is important. I live with people who are very supportive and all that stuff, but it's in a typical American suburb, so you may not even know… I don't even know some of the neighbors. Here, you can't not know people. Here, you want to know about what other people are doing, and maybe you don't sometimes, but everybody's got different ideas about how to relate and what they're doing, and you hear about this and you hear about that, and that's always fun.

References

Colloingridge, D. S., & Gantt, E. (2008). The quality of qualitative research. *American Journal of Medical Quality, 23*(5), 389–395.

McMahan, I. (2015, April 22). Running into old age: A growing number of seniors are completing marathons and triathlons, shedding new light on how exercise affects the elderly body. *The Atlantic*.

Onishi, N. (November 30, 2017). A generation in Japan faces a lonely death. *New York Times*. Retrieved from https://www.nytimes.com/2017/11/30/world/asia/japan-lonely-deaths-the-end.html.

Rhee, T. G., Westberg, S. M., & Harris, I. M. (2018). Use of complementary and alternative medicine in older adults with diabetes. *Diabetes Care, 41*(6), e95–e96.

Rosenbloom, C., & Murray, B. (2018). *Food and fitness after fifty: Eat well, move well, be well.* Chicago, IL: Academic of Nutrition and Dietetics, Eat Right Press.

Chapter 2
Senior Cohousing—History and Theory

> We decided on cohousing; we decided that what we really wanted was to be good to the earth; we wanted spirituality; we wanted mutual support…And I think community's worth it. It's not easy, but it's worth it. …And I think it's the human condition - what we want is some belonging, and support, and acceptance. (Stephanie)

2.1 History

The first modern cohousing community was developed in Denmark just outside of Copenhagen in 1972. Twenty-seven families who desired a greater sense of community and collaboration than found in typical neighborhoods of the time came together to develop a fresh approach to housing (McCamant & Durrett, 1988). With the guiding principles of community and cooperation in mind, these families developed the physical characteristics and the governing structure for their new community that have now become hallmarks of the modern cohousing movement. Architectural features such as community kitchens, communal play areas for children, and common gardens and courtyards served to heighten residents' natural interactions with one another. Shared responsibility for the functioning and upkeep of their community further fostered interpersonal engagement. What resulted was the formation of a more close-knit "neighborhood" without families having to forfeit living in individual homes (Bamford, 2005). In a manner, it was an attempt to re-gain what was thought to have been lost through modernization—a hearkening back to village living where community residents worked together to maintain their way of life.

Senior cohousing, then, was an adaptation of the modern cohousing movement applied to older adults. The first senior cohousing community, called Midgården, was established in Denmark in 1987 (Durrett, 2009). This community was pioneered by two women, Tove Duvå and Lissy Lund Hansen, who championed independent housing for older adults (Durrett, 2009). As their model, they used the already estab-

S. Cummings and N. P. Kropf, *Senior Cohousing*, SpringerBriefs in Aging,
https://doi.org/10.1007/978-3-030-25362-2_2

lished intergenerational cohousing communities in Scandinavia. The community opened with 9 single older women as the initial residents.

The idea of cohousing for older adults caught on quickly in Denmark; but, the reality of working with the government to sponsor such senor cohousing, learning the needed skills to work with architects and developers, and creating a beginning community that was able to effectively confront and solve problems was daunting to many.

Then, in 1995 Henry Nielsen created a comprehensive model to help guide cohousing-interested seniors in the often confusing and challenging process of creating a cohousing community. Nielsen gained his expertise in the field of older adult cohousing through his work with the Danish nonprofit Quality of Living in Focus. His model addressed issues related to design, community size, aging-in-place and member participation. He stressed that the participatory process is essential to the creation of a strong and vibrant cohousing community, as it involves the members in the construction and design of their future community. Nielsen's model describes a step by step process to follow, consisting of 2 phases and three study groups, in order to develop a cohousing community. Phase one is called the "feasibility phase," and is identified with discerning whether or not a senior cohousing community is possible in a given area, finding a site, and exploring how the community will be financed. Phase two is called the "information phase." This phase involves locating other seniors interested in participating in a senior cohousing community, along with honing-in more on how development will take place. Following the "information phase" is a series of three "study groups" that cover specific topics: aging successfully, participatory design, and policy (Durrett, 2009). Furthermore, Nielsen's model identified a number of key players, in addition to future residents, who are essential to the developmental process. These include a third party advisor and project manager who work to coordinate between all the people involved; local officials who help with zoning, public services availability, etc.; a developer who works with the residents and oversees the project's development and financial process; and an architect who designs the community in consultation with the members. While the process of developing a cohousing community can be an arduous one, Nielsen's model helped pave the way for many by demystifying the complex activity of senior cohousing development. As a result, the number of senior cohousing communities in Denmark and beyond began to grow. By 2009 there were 2,800 senior cohousing units in Sweden and 2,100 in the Netherlands (Glass, 2009), while by 2015 there were 250 senior cohousing communities in Denmark alone (Penderson, 2015).

2.2 Cohousing in the U.S.

Moving to the American context, McCamant and Durrett (1994) are credited with coining the term cohousing and introducing the concept to Americans in the 1970s (Glass, 2009). The two were architecture students on a year abroad at the University of Copenhagen. During his daily commute to the university, Charles Durrett came across a newly developed cohousing community. Intrigued, by the lively social interaction on display, Durrett and McCamant decided to learn more about cohousing. They spent 14 months in the early 1980s visiting close to 200 cohousing communities, studying many in detail and even living in a few to gain a deeper understanding of the structure, principles and workings of cohousing. After returning to the U.S., they self-published a book on the topic in 1988 and three years later built the first cohousing community in the U.S. fashioned after the Denmark cohousing model (Verde, 2018). Since the early 1990s, the number of cohousing communities has grown throughout the U.S. Although it can be very difficult to determine the exact number of cohousing communities, according to the Cohousing Association of the United States (2019), there are currently 165 cohousing developments in 36 states. Of course, these represent only those communities that have chosen to join this cohousing association. The vast majority of these cohousing communities are intergenerational.

Communes versus cohousing. In the U.S., the term cohousing is often confused with "commune". Many of the older adults who we interviewed reported that their friends and family members replied, "What! You've joined a commune?!, when they told them that they had purchased a house/unit in a senior cohousing community. In many peoples' minds vivid images of the 1960s and 1970s "hippie" communes still endure. It is true that communes continue to exist in the U.S. and across the globe. While some hippie-type of communes do remain, the term commune is now used to designate a community in which "most everything is shared" and which may consist of shared housing, co-householding and/or co-living (multiple individuals sharing a house) (Fellowship for Intentional Community, 2019). Communes and cohousing are different in several important ways. In communes there is often: (1) a common purse; (2) a focus on the group rather than on the nuclear family; and (3) 100% income sharing. In cohousing some land and space (e.g. the common house) are co-owned. However, cohousing community residents own their own homes and can sell them on the open market. In addition, while community members may have close relationships, there is no desire nor expectation for the group to become the primary focus. Rather, it is understood that the main emotional relationships remain within the family/unit. Lastly, in cohousing there is not a common enterprise nor income sharing. Cohousing, therefore, is more of a tightknit village that shares responsibility for common areas and promotes strong community while maintaining the concepts of individual income, property, and households.

Senior cohousing in the U.S.. Given the quick spread of cohousing across the ocean to America, it would be reasonable to assume that senior cohousing caught on rather quickly. However, this was not the case. Cohousing, focused on those 55 years and older, did not begin in the U.S. until the early 2000s (Glass, 2013). The publication of *The Senior Cohousing Handbook: A Community Approach to Independent Living* by Charles Durrett (2005) helped spur the growth of senior cohousing. In addition to introducing the idea of senior cohousing and describing aspects of the architecture, Durrett clearly described the *process* of creating a senior cohousing community. According to Durrett (2005, 2009) Study Group I is especially important. It is during this 10-week process that potential members look at, and hopefully come to accept,

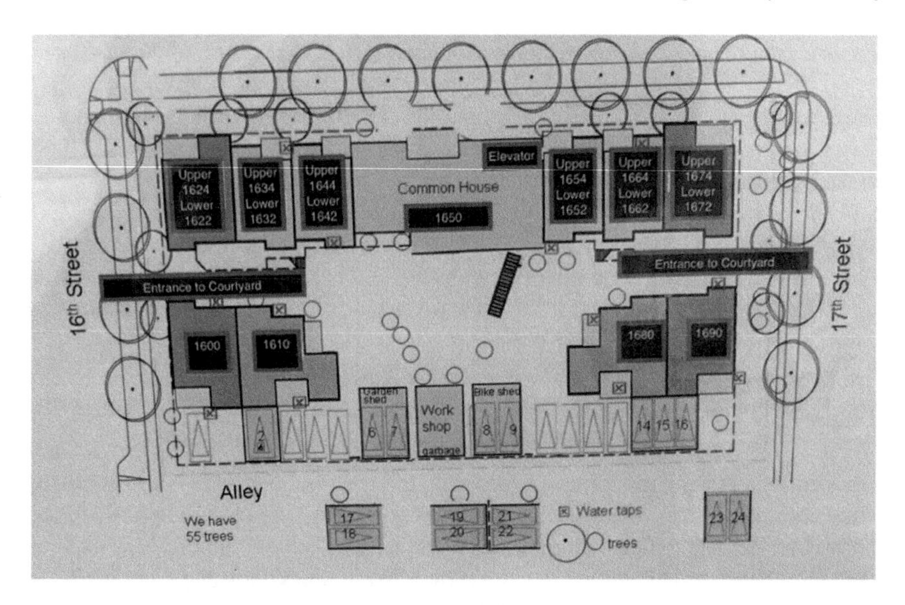

Fig. 2.1 Blueprints for Silver Sage Senior cohousing

their own aging process, examine what it means to them to "age-in-place" and decide how they would like to do this. This type of sharing also enables potential members to learn essential listening skills, get to know one another better and build beginning community bonds. Approximately 40% of those who go through this process decide to forge on and participate in the creation of a senior cohousing community (Durrett, 2009) (Figs. 2.1, 2.2 and 2.3).

Between 2005 and 2007 three separate senior cohousing communities opened in California, Virginia, and Colorado, all which were visited by the authors. The first cohousing community in the U.S. was Glacier Circle in Davis, California. Much like the first elderly cohousing development in Scandinavia, Glacier Circle was conceived of by a single person, Ellen Coppock, who had reservations about moving into a traditional senior housing residence. Concerns about loneliness and isolation were the impetuses that drove Ellen Coppock and her friends, many of whom had known one another since their children were in pre-school, to begin considering options other than the local Life Plan Community. In their own way, they informally engaged in the process of Study Group I. At the end of this process, they decided to move forward and develop a senior cohousing community. They visited an intergenerational cohousing community in California and decided to base their community on this model. Glacier Circle was privately developed by a core group of about eight people who were committed to creating a community of older adults dedicated to living in relationship as they aged. The group first came together in 2002. Although they initially faced many setbacks, once they found a plot of land where they could build and an architect who was willing to work with them as they wanted, the project moved forward, and after several years of hard work, the community officially opened in 2005.

Fig. 2.2 Working toward a Senior cohousing mission statement

The second senior cohousing community to develop was ElderSpirit in rural, Virginia. After looking for land, a group of interested older individuals formed a corporation to purchase the property and received funding from the Research Retirement Fund for predevelopment activities. This group consulted with Charles Durrett and hired an architect. Members of this emerging community served in the role of program manager, who oversaw the physical development and worked with government housing agencies, and in the role of community coordinator, who publicized and recruited members for the developing community. ElderSpirit opened in 2006. The third senior cohousing community to open was SilverSage in Boulder, Colorado. This group was initially made up of aging members of an existing intergenerational cohousing community, Neyland Cohousing, who were exploring a cohousing option for themselves that was more focused on their needs and interests as older adults. They were soon joined by others and found land across the street from an already existing intergenerational cohousing community, Wild Sage. They worked with a non-profit agency and a developer to create the first mixed income senior cohousing community. This group also hired McMamant and Durrett as their architects to design, with community members input, the physical structure and space. This was also the first group that officially went through the process of Study Group I. They opened their doors in 2007.

Fig. 2.3 Blueprints come to life in common spaces

Since 2007, the number of senior cohousing communities has slowly but steadily grown. There are now seventeen such communities listed in the directory of the Cohousing Association of the United States (2019), and 28 are currently in formation or under construction. Similar to the original three projects, these new communities are spread in states throughout the country, range in size, are situated in diverse geographic settings, and possess mission statements with a variety of foci from mutual support and green living to spirituality and community engagement. Likewise, new groups are now employing a variety of developers and architects, including senior cohousing specialists, local professionals and community members themselves, and using a creative array of financial mechanisms to develop and build their communities.

2.3 Theory

The Gerotranscendence Theory of Aging helped shape the framework we used to conduct our study of the lived experience of senior cohousing residents. In particular, this theory was used to guide the development of our structured interview questions and to organize our data analysis and data interpretation, that is, the way we thought about and developed themes from what the residents told us about their lives in their cohousing communities. Gerotranscendence is a developmental theory of aging that posits the continued possibility of personal growth as one ages. It suggests that growth can be seen on three different levels—cosmic, self, social/personal. So, as one grows older, it is possible to "transcend" or move beyond previous understandings and gain new perspectives on fundamental existential issues, the meaning and importance of relationships, and definitions of self (Tornstam, 2005, 2011).

2.3.1 Cosmic Level

- Increased feeling of oneness with the universe
- Increased connection to both past and future generations
- Increased ability to reflect on and incorporate earlier aspects of life
- Increased acceptance of the mystery of life; accepting that not all things can be understood or explained from a rational point of view
- Increased ability to see the universal in individual objects
- Decreased fear of death.

2.3.2 Self Level

- Increased self-acceptance
- Increased altruism
- Increased integration of different pieces (both good and bad) of one's life
- Increased sense of wholeness and coherence
- Decreased self-centeredness
- Decreased obsession with the body.

2.3.3 Social/Personal

- Increased desire for solitude and meditation
- Increased acceptance of other ways of being and doing things
- Decreased interest in superficial relationships
- Decreased need to conform to social roles and norms
- Decreased attachment to material possessions
- Decreased judgmentalism.

The authors went into this project believing that growth can occur at all ages, and saw in the theory of gerotranscendence a framework for looking at different aspects of possible growth as one ages. What is presented in this book should not be viewed as an effort to support or critique the theory of gerotranscendence. Rather, it is an attempt to explore whether the experiences described to us by senior cohousers would reflect any of the themes noted above and, if so, would the participants attribute some of their growth experiences to life within cohousing rather than to aging itself? In other words, might the experience of living in senior cohousing help foster gerotranscendence? In the next chapter we turn to a description of the communities we visited, and an explanation of the shared governance and communal management processes used. These set the scene for the remaining chapters and further exploration of the lived experiences of community members and the impact of these experiences on their growth and development.

References

Bamford, G. (2005). Cohousing for older people: Housing innovation in the Netherlands and Denmark. *Australian Journal on Ageing, 24*(1), 44–46.

Cohousing Association of the United States (2019). https://www.cohousing.org/. Accessed 1 February 2019.

Durrett, C. (2005). *Senior cohousing handbook: A community approach to independent living* (1st ed.). Berkeley, CA: Habitat Press.

Durrett, C. (2009). *Senior cohousing handbook: A community approach to independent living* (2nd ed.). Gabriola Island, BC, Canada: New Society Publishers.

Fellowship for Intentional Community. https://www.ic.org/. Accessed 17 February 2019.

Glass, A. P. (2009). Aging in a community of mutual support: The emergence of an elder intentional cohousing community in the United States. *Journal of Housing for the Elderly, 23,* 283–303.

Glass, A. P. (2013). A conceptual model for aging better together intentionally. *Jounal of Aging Studies, 27,* 428–442.

McMamant, K., & Durrett, C. (1988). *Cohousingg: A contemporary approach to housing ourselves.* Berkeley, CA: Habitat Press/Ten Speed Press.

McMamant, K. & Durrett, C. (1994). *Cohousing: A contemporary approach to housing ourselves* (2nd ed.). Berkeley, CA: Ten Speed Press.

Pendersen, M. (2015). Senior cohousing communities in Denmark. *Journal of Housing for the Elderly, 29,* 126–145.

Tornstam, L. (2005). *Gerotranscendence: A development theory of positive aging.* New York: Springer Publishing.

Tornstam, L. (2011). Maturing into gerotranscendence. *The Journal of Transpersonal Psychology, 43*(2), 166–180.

Verde, T. (January 20, 2018). There's Community and Consensus. But It's No Commune. *New York Times.* https://www.nytimes.com/2018/01/20/business/cohousing-communities.html. Accessed 10 February 2019.

Chapter 3
Communities Visited—Overview

> We made a commitment to be in community. That both has to do with running the place and all of that entails but also in terms of being there for each other.... (Bob)

In the U.S., there are 17 senior cohousing communities that are currently up and running. The first was established in 2005 and the last came onboard in 2018. To achieve a good representative understanding of these communities, we visited 12 communities that ranged in size, geographic location, number of residents, types of structure (homes or condo units) and urban/rural/suburban setting.

An integral aspect of senior cohousing is the active participation of the older residents in the design and development of the community. All but one of the communities that we visited was developed by an original group of older residents who were strongly committed to the cohousing principles noted in chapter one. Most frequently, these original members reached out to a developer and/or architect early in the process to help guide them thought the design development and construction phases. One team in particular, Charles Durrett and Kate McMamant, who are leaders in the senior cohousing movement, provided consultation to several of the senior cohousing communities. Others recruited local architects and contractors while a member of one smaller community, who had a background in real estate and construction, oversaw most of the process herself.

3.1 Cohousing Communities

Following is a brief description of each of the communities we visited and an overview the governance, decision-making and recruitment processes they employed.

S. Cummings and N. P. Kropf, *Senior Cohousing*, SpringerBriefs in Aging,
https://doi.org/10.1007/978-3-030-25362-2_3

Fig. 3.1 Acequia Jardin

3.2 Alcequia Jardin

Established in 2013, Alcequia Jardin is a small community set on 1.1 acres located just a few miles away from Old Town Albuquerque and next to a food co-op. Nestled off of a main road, the community consists of the 5 duplexes (10 homes), a small community room and a guest-suite used for short-term visits by family and guests. The community prides itself on being environmentally and socially responsible. Homes, which are energy efficient and range in size from 800 to 1200 sq feet, are gathered around a community garden and courtyard maintained by the members. While members have organized a weekly book club, occasionally share dinners and engage in other activities that take place on-site, they are also very active in the larger community (Fig. 3.1).

3.3 Elderberry

Elderberry is located in a rural area about 30 min north of Durham, North Carolina. Started in 2011, Elderberry is adjacent to a multi-generational cohousing community named Potluck Farms. With strong principles rooted in ecology and sustainability, the residential units are energy efficient and a solar farm is being added. There are multiple organic gardens and a chicken coop for eggs. The 18 home units have a limit of 1200 sq feet and are either duplexes or quadplexes. There is a common house with a kitchen, library, and meditation area. While the members of Elderberry have organized several activities and events for themselves, they also interact with those living in Potluck Farms.

3.4 Elder Family Fellowship

Elder Family Fellowship (EFF) is located in the rural community of Whittier, North Carolina. As a senior cohousing community, it is the smallest of the communities visited with just two current residents. The goal is to recruit additional members to become part of the "family" which is the preferred term over "residents". EFF is part of a larger community, Union Acres, that was started in 1989. One of the founders is now living in EFF, which is physically sited within the Union Acres property. Whittier is in a beautiful area located in the western part of the state in the Smokey Mountains. EFF consists of a single building with multiple floors, separate units, a large kitchen and a great deal of privacy. All other homes in Union Acres are spread over about 90 acres. There is a communal building where residents of both EFF and Union Acres gather for meetings and social events.

3.5 Elderspirit

Elderspirit is located in Abingdon, Virginia which is a small town in the southwest part of the state. Started in 2006, the community has a strong spiritual foundation and includes Christians, Jews, and Buddhists. In addition to single dwellings, there are rooms in the communal building that are rented to make living there more affordable for those in lower income brackets. About 25 people currently live in Elderspirit. The property is integrated into the small town, and it is easy to walk to the downtown area which has shops, a community theater, and restaurants. In addition, the property borders on the Virginia Creeper Trail, a 34 rail-to-trail system for walking, biking and recreational uses.

3.6 Glacier Circle

Glacier Circle, begun in 2005, was the first senior co-housing community established in the U.S. Organized by a group of friends with common roots in the Unitarian Church, this community is located in suburban Davis, California. The eight townhouses, 1,000–1,400 sq feet in size, and a spacious community house encircle a mature and well-maintained garden consisting of colorful flowers and trees. While the front of the community faces a neighborhood consisting of single-family dwellings, the back gate opens onto a greenway and reserve. Because current members have aged-in-place and are older (current average age is 90 years), they have chosen to jointly hire outside help to handle the community's finances, maintain the garden, and to cook communal dinners 4–5 times per week.

3.7 Mountain View

This cohousing community is located in the city of Mountain View, CA, which lies in the heart of Silicon Valley and is situated on the southern end of the San Francisco Bay Area. Mountain View is a small high-end city with a walkable downtown. The senior cohousing community at Mountain View began in 2014. One structure houses the 19 condo homes, which extend from 1350 to 2090 sq feet and are situated along three floors. Front doors and back walls contain windows that allow in ample sunlight and foster easy socialization. A spacious community area consisting of a large kitchen, meeting space and a mail area is positioned in the center of the first floor. Residents maintain a large communal garden that includes a wide variety of flowers, trees and vegetables. Guests and family members are welcomed to stay at the "farmhouse", a separate adjacent building that houses two well-appointed bedrooms, a kitchen and a living area.

3.8 Phoenix Commons

Phoenix Commons is located in a busy urban area of Oakland, California known as Jingletown. This 41-unit condo complex, begun in 2016, is the largest senior cohousing community we visited. It is also the most diverse community explored in terms of backgrounds, race, sexual orientation and work status. The units in this airy multiple-floor building face one another and are separated by open air walkways. This construction grants privacy while encouraging residents to mingle in outdoor areas. A spacious common area, located on the first floor, contains a large kitchen, an exercise room, a library and gathering/television area where residents meet to visit and cheer on their favorite teams. Phoenix Commons is positioned along the Oakland waterfront and at the base of a walkable bridge to Alameda Island.

3.9 Sand River

Situated a few miles from downtown Santa Fe, Sand River is a small cohousing community positioned on 3.5 acres of land. The community, which is celebrating its 10th anniversary this year, consists of 14 adobe style duplexes with 28 units and a common house that contains a large kitchen, a small library and an office area. The duplexes and community house are joined by gravel walkways while natural southwestern-style plantings and grasses line many of the buildings. The cohousing community rests about a quarter mile off of a main road that runs through Santa Fe. A large gym, commercial establishments and restaurants are a quick drive away. Sand River prides itself on being a LEED community with sustainable design features that save resources and promote renewable, clean energy.

3.10 Sarah's Circle

Sarah's Circle is a non-profit agency and apartment community located in the heart of the Adams-Morgan neighborhood of Washington D.C. Begun in 1983, Sarah's Circle was established to provide affordable housing and a vibrant community for very low-income older adults. Although not a traditional senior cohousing community in that it was not designed and developed by the older residents themselves, it is included in this book as an example of older adult community living that is accessible for those with very limited financial resources. The three-story building houses 36 studio, one-bedroom and 2-bedroom apartments and is anchored by a large community center in which residents gather to socialize and participate in activities such as weekly creative writing and watercolor classes. One of the residents functions as the manager for the building while other residents serve on the Sarah's Circle board of directors. Sarah's Circle is supported by HUD-funding and donations. Residents contribute one-third of their income toward their monthly rent (Fig. 3.2).

Fig. 3.2 Sarah's Circle

3.11 Silver Sage

Silver Sage is situated in North Boulder (NOBO), an artsy part of Boulder, Colorado. The community is part of a high-density living environment; for example, Wild Sage, a multigenerational cohousing unit, is located across the street. The physical layout of Silver Sage is extremely attractive with a communal garden containing flowering plants, vegetables and herbs. There are sixteen units in total, with a subset of "affordables" that are set below market value. This arrangement was created to secure a tax benefit from Boulder when Silver Sage was built. To bring down communal property costs, Silver Sage rents out some space to local artists. The Silver Sage residents share other space including a workshop with powertools, an art room, exercise room and a meditation room. The community is located about 3 miles from downtown where the University of Colorado, Boulder is located. Down the street is an accessible public transit bus system that many residents use to travel downtown. Around NOBO are restaurants, paved running/walking trail, biking trails. As one resident said, "everyone here has a Subaru and a bike."

3.12 Walnut Commons

Walnut Commons was originally designed as an inter-generational cohousing community, and while members actively encourage residence by younger adults, no common space is allocated on property for children. As a result, residents mainly fall in the 55 year + age group. This community, which is located in the heart of the extremely popular city of Santa Cruz, was opened in 2014 and consists of 19 units situated in a three-floor condominium building. In addition to the condo units, Walnut Commons houses a common area containing a kitchen and two co-joined meeting spaces. The common area opens onto an outside patio graced by flowers and plants. Walnut Commons encourages environmentally sustainable living. Because of its location almost everything that anyone could want—a public library, museums, restaurants and shops, a movie theater, grocery stores and the beach—are within easy walking distance.

3.13 Wolf Creek Lodge

Wolf Creek Lodge, which opened its doors in 2009, is situated in Grass Valley, California, a small historic gold rush town that is now home to numerous lively entertainment events, art galleries, restaurants, and a theater. It is located in the scenic western foothills of the Sierra Nevada Mountain range. Wolf Creek Lodge is an airy and light-filled three-story building that contains 30 individual condo homes, 4.000 sq feet of shared indoor common space, a large outdoor patio where meals

Fig. 3.3 Wolf Creek Lodge

are often shared and a spacious well-maintained garden. The common area, located on the first floor provides space for a large kitchen, two guest bedrooms, laundry facilities and both large and small meeting areas. Most of the units have patios and all open onto a shared walkway. Grocery stores, restaurants, the post office, a hiking trail and many other amenities are within easy walking distance. Residents engage in many shared activities within the lodge itself but are, also, very active in the town and in area political activities (Fig. 3.3).

3.14 Governance and Management

In addition to the cohousing principles of neighborhood design and common facilities described above, each senior intentional community that we visited actively engaged in resident management. This means that all the residents share in both the decision-making and running of the community.

3.15 Decision-Making

All communities, whether large or small, typically hold monthly residents' meetings in which proposals concerning the building, grounds and community policies are discussed and votes are taken. One of the most distinguishing characteristics of cohousing is the use of consensus decision-making. This means that all residents are encouraged to actively participate in discussions and to vote on proposals. How cohousing groups implement consensus decision-making does vary, however, and is the topic of much discussion among senior and intergenerational communities alike.

The decision-making process in the larger senior cohousing groups we visited tended to be more structured. In some, prior to bringing a topic to larger monthly meetings, members are encouraged to form exploratory groups that develop a draft proposal, receive feedback from key, or all, members of the community and make suggested revisions to the proposal before bringing it forward for discussion at a monthly meeting. Alternately, other communities have executive committees that view all proposals and request additional information/clarification, as needed, prior to the discussion of the proposal at the large group meeting. Two of the communities we visited also had an additional process in which, proposals that gained community interest but failed to secure an affirmative vote, were referred back to a group of interested residents for continued discussion, research and, perhaps, re-introduction at a future meeting. In smaller communities, proposals are often introduced with less formality during the monthly group meeting and then discussed until all residents agree to accept, further investigate or drop the idea altogether.

Regardless of the process, decisions are typically made by consensus, although what is meant by "consensus" varies by community. To some, this means that 100% of the residents must agree in order for a decision to be implemented. Therefore, one person could theoretically hold up or put an end to an idea. One community that did hold to this definition of consensus also had a proviso requiring that all "no" votes to a proposal be based on consideration of the community's best interest rather than on personal preferences. A few of the larger communities also followed a structured process to help order the large group discussions and votes. Under this structure, a vote is taken after a proposal is introduced and discussed. Residents then hold up one of three differently colored cards to indicate their votes of agree, disagree or abstain. Proposals failing to achieve consensus are then either discarded or tabled for further exploration and discussion. As stated above, some cohousing groups, both large and small, require proposals to have 100% support while others believe that consensus has been achieved when 80% of the residents agree.

3.16 Management

Member participation forms the heart of cohousing communities. In most cases, all the work required to maintain the buildings and grounds, and to operate the community is accomplished by the members themselves. When new individuals buy a property, they realize that they are *also joining a community* and must share in these responsibilities. The work of the community is mainly completed through teams or committees. Rules such as participating on two committees or working 12 hours per month are common. Residents join teams reflective of their skills and interests and the community's needs. Common teams include gardening, maintenance, membership, finance, and dinning.

The amount of work required depends upon the size of the community as well as the skills and interests of those involved. Some communities, for example, have extensive landscaping and flourishing gardens filled with flowers, trees and vegetables due to the members' love of horticulture. Other communities have lovely, yet more modest, gardens that reflect their members' interest and time. Likewise, joint dinners may be more extravagant affairs or potlucks depending upon the dinner team's, and community members', interests. One smaller cohousing group decided to forgo preparing community dinners all together and instead agreed to jointly hire a chef to cook once a week and to engage in occasional "pop-up" meals. Residents who prefer more solidary participation may also engage in activities such as newsletter development and website maintenance or serve as the community treasurer.

Cohousing communities have diverse ways of accounting for members' participation. One of the most structured methods we encountered required that all members report their monthly activities to a volunteer who kept an account of everyone's hours and notified members if they were falling behind. On the other end of the spectrum, there were several communities that followed the "honor" system and allowed members to keep their own record of the hours they contributed, all the while knowing that certain members would consistently fall short. Several communities discussed the possibility of initiating a "play or pay" system in which members who either chose or were not able to fulfill their participatory commitment would pay higher monthly fees. At the time of our visits, however, none of the groups we interviewed had adopted this strategy.

3.17 Member Recruitment and Vetting

New members are recruited during several periods throughout the life of a cohousing community. The first is when a new cohousing community is beginning to form. Those initiating the formation are often a handful of individuals who are searching for like-minded people to join them in the creation of a vision for the community and in the actual development of the community itself. Recruitment of interested members at this nascent stage may involve word of mouth, website notices and/or newspaper and radio announcements. Founding members often spend countless hours spread over several years developing architectural plans, searching for property, securing funding, overseeing construction and developing initial bylaws and policies before the community opens its doors. Members involved at this stage often comment on both the tremendous energy required and the depth of relationships formed while involved in this process.

The recruitment of owners to purchase homes after construction has begun often takes place using the same methods as described above. Once all units have been sold and occupied, new units become available when current members relocate or die. Recruitment of new members takes many forms depending upon the particular community. Larger senior co-housing communities often have active websites that post information concerning available homes. Most have a contact person or membership coordinator listed on their website who responds to requests for information. Regardless of the current availability of a unit, interested individuals are often invited to sign-up for a community's monthly newsletters or to schedule a visit. The goal is to educate those making inquiries about the realities of senior cohousing and about the nature of the specific community, in particular. Communities with limited members often do not have the resources to sponsor an on-going newsletter or an active website that lists available properties. In such cases, owners may look to sell their homes on their own or via a real estate agent. Because members own their property, they, or their heirs, may sell their home to whomever they wish. This reality has, at times, resulted in a new owner who was not aware of, or interested in, cohousing mistakenly purchasing a home within a cohousing complex. Although a rare event, membership teams do all in their power to avoid this potentiality by educating prospective buyers about the mission and nature of their communities. One group we visited retains the right of first refusal so that, in cases where a potential buyer is not well-suited for the community, the group may purchase the property and sell it on their own to someone who is a better fit with the mission and workings of the group. Another community actively recruits new owners through Unitarian churches when a home becomes available. In the end, all cohousing communities seek to recruit new members who not only understand cohousing principles but who also aspire to live, and actively participate, in community.

Chapter 4
Why Senior Cohousing?

> It seemed like a wonderful idea to grow old together… especially to grow old together with friends. (Lily)

What draws people to move in later life? What rationale do they have for leaving behind homes and neighborhoods where they have often lived for decades to set out on a new adventure? An initial step in making this decision is to sort through the all the residential situations that are available, learn their features and make a determination of what you are truly seeking at this point in your life.

Living options for older adults who are active and independent include many alternatives. Large master-planned retirement communities abound. Some, such as The Villages in Florida or Sun City in Arizona, are home to as many as 115,000 adults 55 years of age and over. As advertised on its website (2018), The Villages is a "Fun and affordable active adult community where everything you could possibly want, need, or dream of doing in your retirement years is just a golf cart ride away.". The layouts and features of these communities are pre-planned by developers, and on-site services and activities are provided by a management company. Another living option is a senior retirement residence. These are much smaller affairs and are often more akin to typical apartment buildings. Retirement residences range from low-income rentals to luxury condos and may include amenities such as recreational activities, hairdresser services and transportation. These are developed and managed by a variety of groups, such as for-profits, non-profit, governmental, and religious organizations.

With the advent of health issues, there is often an accompanying need to have additional supports in place. Life Plan Communities (previously known as Continuing Care Retirement Communities) are also another option. Life Plan Communities include apartments or houses for independent living as well as Assisted Living and Skilled Nursing Home care for those who require more assistance as they age. Residents often pay a sizeable entry fee for admittance into a Life Plan Community when they are able to live independently. However, if their cognitive and/or physical functioning deteriorate over time they are able to move into on-site assisted living

S. Cummings and N. P. Kropf, *Senior Cohousing*, SpringerBriefs in Aging,
https://doi.org/10.1007/978-3-030-25362-2_4

or skill nursing units. All residents pay monthly fees to cover meal, recreational, management and other services.

With many independent living options available to older adults, why do some actively choose senior cohousing communities? What aspects of senior cohousing lead individuals to this type of living arrangement? From our interviews, both *pushes and pulls* drive people to first investigate and then actively choose to live in senior co-housing communities. *Pushes* are those factors that create dissatisfaction with a current living situation. On the other hand, *pulls* are those reasons that induce people to select a SCC. Both types were discussed by the people that we interviewed.

4.1 Pushes: The Desire to Leave a Current Residence

Aging-in-place. As far as *pushes*, many individuals first came to the understanding that their current living situation was not optimal for aging-in-place by watching the aging process of parents and friends. Respondents reported having witnessed the creeping isolation that often invades someone's life due to growing functional impairment or increased frailty. Having problems driving in the dark due to poor night vision or losing a driver's license all together, experiencing difficulties negotiating the front steps, and having problems travelling due to fatigue can lead older individuals to become isolated within the very homes that had once provided them great comfort. One woman stated "I watched my mom as she stopped driving …And when she stopped driving, everything tanked pretty quickly. Emotionally, physically and mentally. So, at that point, I looked at where I was living and thought, "What happens if I stop driving…and I thought, that is just not sustainable, I don't want to be isolated in my own home as I get older…" (Mary). A resident of another community pointed to a similar experience she had "I have two friends who have steps, …She broke her leg and they can't maneuver around their place. They had to put stair lifts ….and then she's isolated when he goes to work. It's horrible". (Samantha). Betty told a comparable story, "My wife and I were taking care of her mother and she was at a local facility, but we would take her out and take her to our [home] periodically, and we realized that at one point she couldn't go up the stairs by herself. We had to help her on both sides sometimes and if the young man hadn't been next door, we would have had trouble getting her down the stairs. And this whole experience, for four years, of taking care of her mother and being in charge… made us realize that we needed a plan and so we started looking at co-housing" (Fig. 4.1).

Fig. 4.1 Navigating steps can be difficult and dangerous

Community disengagement. Another factor that led others to the decision to leave behind their current homes and seek out a living situation that promised greater engagement and interaction was the realization that they no longer knew their neighbors or felt engaged within their communities. When individuals are employed, they are often busy and socially occupied at work. As a result, they cherish their homes as places of refuge. However, after retirement some look around and realize they don't know those who live next door to them anymore. Long-time neighbors may move away, communities can change and, at times, areas of refuge may begin to feel more claustrophobic and isolating. Regardless of type of locale—rural, suburban or urban—this is a common reason that residents report moving to a SCC. One individual who lived in a major city reported, "…and I had been living in the apartment. I was there for 10 years, and I didn't know my neighbors." (Rita). One woman who had lived in the same neighborhood for decades noted that her neighbors typically commuted an hour each way to work, picked up their children from school, came home, went inside their houses and didn't come out again. On the weekends those living nearby were busy taking their children to activities or doing errands. "And I … we only knew one set of neighbors and we just knew them just to say, "Hi! How are you?" We never really did anything with them. I mean, we could have dropped dead inside the house and nobody would have known…" (Paige). Another suburbanite stated "A lot of times in the suburbs we didn't, well we recognized people but we didn't really know people and didn't necessarily socialize" (Dottie). Those living in rural areas often face even greater challenges, "We lived on the top of a hill. We had eight acres. We literally had never met our neighbors" (Kate).

4.2 Pulls: The Move to a SCC

Concepts of community. One of the major lures to life in senor cohousing is the sense of a close-knit community, a place where one is known, knows others and can easily interact and engage with those who share similar values and interests. During our interviews several residents referred to senior co-housing communities as "The new old-fashioned neighborhood". As one senior cohouser who grew up in Philadelphia put it, "And I really wanted to have a community in which there was a feeling of family, which I had as a child growing up…. We knew everyone in the block, you know? We played together, we spent time together…" (Joy). A woman who had been raised in small town America shared a similar sentiment, "So I grew up in a tiny little town, I knew everybody in the town… There were 3000 people in the town. And people I went to preschool with I graduated from high school with. So, I had this little tight community" (Fig. 4.2).

While not all members of senior cohousing grew up in close knit neighborhoods, many did have previous experience with some type of community living which they had enjoyed and found rewarding—from being in the Peace Corps, to living in a college dorm or with a group of friends after graduation, being a member of a spiritual community or from a culture that emphasizes close personal relations, "…in the Pilipino culture everyone is expected to be social and share. So growing up I saw my mother inviting friends over, filling the house…" (Chris).

Others, however, has no such experiences but still longed for a place to live where they could connect deeply and easily with others. "So, I really loved the idea of being together, doing things together, and we could be as happy as we could be, you know…I really had no misgivings at all. My husband was a not very social person.

Fig. 4.2 Vision statement—Elder Family Fellowship/Union Acres

To me it was an extra reason to like the idea of being with other people and doing things together." (Lily).

Many individuals also stated that while they had no previous experience living in community, they were looking and hoping for a living situation that would allow them to develop relationships like the ones they had experienced in their own families or with very close friends. "And I really wanted to have a community in which there was a feeling of family, which I had as a child growing up." (Ava). When considering cohousing, most residents seemed to be realistic, however. As with many family situations, they realized that there would be some individuals with whom they would connect more easily and others who would be more challenging for them. "There's some people that you're going to have problems with, it happens in any place, and other people who just open their heart right up to you, and it's extra family in your heart." (Caroline). Andie also expressed similar realistic expectations "I knew with 41 units, I knew that there were going to be people that were going to be my best friends. I mean, just odds are, and I knew that there was going to be people that were not going to be my best friends."

Common interests and values. Those we interviewed reported they had anticipated that others who lived in senior cohousing would share their interests and values. This assumption makes sense, as SCCs are typically founded upon a set of principles that are primary to cohousing communities (see Chap. 1). For this reason, their expectations of developing close friendships and being able to maintain healthy relationships, even with those with whom they weren't close, seemed well-founded. "But I was just taken with the fact that this is an intentional community, that people have made the decision that they had wanted to come here and share their lives with one another, and also by the values, on the assumption that everyone who decided to come was sort of subscribing to one degree or another, to those values..." (Lucy). In addition to sharing similar values, those living in senior cohousing tend to be more liberal in their politics and cultural orientations. On average, senior cohousers are well educated, well-traveled and open to diversity. "I have always found community to be tremendously important...So that was the main thing for me. I was attracted to the idea of diversity, diversity of race, income, sexual orientation, any of that kind of diversity you can think of. That is one of the values here, and it's not always easy to attain, but I found that tremendously important." (Maggie). Dixie also reported "I think we believed politically they were... people we would be compatible with. [This town] is very liberal in its freedom to be whatever you want to be, religious, even politically. There's a lot of tolerance for that."

Another value highlighted in the mission statements of many SCCs is that of respect for the environment, and residents often cited this as a major draw for them. They spoke of the desire to own less and downsize, to live in ecologically friendly structures and to care for the earth. When speaking of what drew him and his wife to their particular intown co-housing community, one resident reported "Well, they're active. They're progressive. They're concerned. There are several very strong environmentalists. And these were the things that we ourselves are active in...." (Chip). Kate, a member of a rural community, also noted "This community just felt really

Fig. 4.3 Preparing for a shared meal

good. We liked the people. We liked the concept of trying to live a greener life, trying to live in community, all those kinds of things, and staying active while you age, was really appealing to us …".

Most units in senior cohousing are smaller than the homes in which residents had previously lived and many individuals and couples saw divesting themselves of unneeded possessions as a benefit. "So, the idea of living in a community, having a smaller footprint, was something else important" (Marge) and "So, we thought this was a good idea to downsize, which we did by 75%" (Jim), were the types of sentiments we often heard. It should be noted that living in a smaller spaces is made more practical in co-housing communities due to the sharing of common areas and equipment. The common house provides an "extra living room" where folks can gather to watch TV or just relax (Fig. 4.3). As Dottie, stated "But the sharing, having people around to do things with, not everybody owning a washing machine, things like that...being more ecologically sound, so sharing equipment and having people around, is the upside of cohousing."

4.3 Why *Senior* Cohousing?

When talking to others about the book that we were writing on senior cohousing many people we spoke to were very excited about the idea of cohousing. Living with others, being able to make and connect with friends easily and having a sense of community—all these features are very attractive to many people. But .... the one question that continued to pop up was, "Why ***senior*** cohousing?" Many people stated, "I don't think I'd want to live with a bunch of old people". Interestingly, when we asked residents if they had had any concerns about moving into a SCC, they often shared the same misgivings about living in a community with "old people". However, after exploring and visiting both intergenerational and senior cohousing communities, those we interviewed found that their views had changed.

While many often found multi-generation communities to be attractive, they also came to the realization that by their very nature such communities must first focus on the welfare of the children. As one resident explained, "...we realized that as much as we like kids and everything, we found out that the intergenerational were not intentionally, but de facto, they had to be controlled by the kids, the needs of the children. So, this just seem to be a better fit for us..." (Dorothy), and "... I have heard that in those types of communities, it's all about the kids, all their meetings and everything... I have grandchildren, they're all grown now, but that just wasn't what we really wanted." (Dixie). We also heard from one man who had moved from a multi-generational to a senior cohousing community, "But there go the connections. In the multi-generational, everybody's working. There are people who love it, but let me tell you that I was alone so much of the time in a multi-generational community." (Pete).

When comparing senior communities with multi-generational, many individuals also quickly perceived the benefits of communities that purposefully included universal-design, which would accommodate their needs should they require more assistance as they aged. Accessible entrances with no steps, an elevator for those living on higher floors, widely spaced door-ways should a wheelchair ever be needed - all promised a greater ability to age-in-place. But, physical space, while very important, is not the only element that enables individuals to age well. Practical and emotional support are also needed. One resident explained, "...a developer frequently said, "80% of what older people need as they get older, they can provide for one another rather than assisted living places." Residents realized that by living with others at their same stage of life, they would be more easily able to talk about their challenges, plans and wishes for aging. They desired a place where they could be open with others and provide mutual support. Stephanie explained that the provision of mutual support is, in fact, one of the most important benefits she found in senior cohousing.

"When I'm at home, here at [SCC] I feel young and active. And when I go out, I get treated like I'm old. And so, there's some vitality here, for seniors, that the culture doesn't accept yet…That's like one thing about being with seniors. And I think the commonality gives a lot of strength. There's a lot of advantages just being with people that are like you."

Reference

Thevillages.com. (November, 17, 2018). Retrieved from https://www.thevillages.com/.

Chapter 5
Staying Active and Engaged

> My social life has greatly improved by living here. Because you just get to see your friends all the time. (Andie)

> My life has taken almost a leap, in sense, joi de vivre, or whatever that is. I don't speak French, but it really, really has. (Jenny)

An aspect of senior cohousing life that draws many people to this type of living arrangement is the opportunity to engage with others. As discussed in earlier chapters, several individuals commented that they knew few people in their previous neighborhoods which led to a sense of disconnection. Residents in senior cohousing desire something different in their lives. As Marie stated, "we're kind of pilgrims on a journey here, redefining what it means to get older in a community." Having people close by promotes a sense of involvement and fosters spontaneous interactions as well.

Within the communities, the housing units are purposefully designed to create connection. In several SCCs, housing or condo units had windows that faced outward toward shared spaces such as the common house or gardens. In this way, residents would see others who were "out and about" thus increasing the probability of social interaction. One woman stated that she really enjoyed looking outside her kitchen window first thing in the morning. She would sit and drink her coffee and watch others in the community starting their day as well. Likewise, SCC homes are intentionally designed with less square footage which encourages residents to mingle in the shared areas. For example, some individual units do not have washers and dryers and most do not have separate mailboxes; so, shared laundry and mail rooms become areas of connection with others. One woman discussed how the common spaces promoted interaction, "the gardening actually has turned out to be quite a social experience and then just wandering out in the evening to pick your lettuce for dinner, for a salad, and running into somebody out there is a lovely experience." (Noreen) (Fig. 5.1).

In addition to the interactions within the cohousing communities, SCC residents are also were very involved with their surrounding community. Often these activities are aligned with the principles or focus of the SCC (e.g., spirituality, ecology,

S. Cummings and N. P. Kropf, *Senior Cohousing*, SpringerBriefs in Aging,
https://doi.org/10.1007/978-3-030-25362-2_5

Fig. 5.1 A gardening team

social action). While many interactions take place within the cohousing area, the surrounding communities greatly benefit from the talents, skills, and dedication from the senior cohousing residents. Although senior cohousing residents have high levels of activity, many described the need to balance social roles with solitude. This "public-private" tension is managed in various ways as individuals determine how and when they will maintain social ties and also claim time for individual needs and space. The remaining sections of this chapter will explore the social connections and activities of residents, and ways that they capture individual time within a communal living arrangement.

5.1 Activity Within Senior Cohousing

One of the push factors for leaving a current home and moving to an SCC is the opportunity for greater engagement with others. As described in Chap. 4, traditional living arrangements can make it challenging to remain involved because of multiple factors such as lack of transportation, distance, or scheduling issues. Within the SCCs, residents build in social time that strengthens bonds and relationships. Additionally, participation in the shared responsibility for maintaining the overall community creates opportunities for residents to work together and to develop networks that increase engagement. As noted above, the design of the communities also increases natural occurring interactions. All of these factors relate to the high level of activity that occur within and beyond the SCC.

Purposeful activities. Purposeful activities are those that are organized and planned to bring individuals together. Community meetings are such a time when community residents gather. In fact, communities have the expectation that residents will prioritize attending these meetings. As Trent states, "…once a month we have a community meeting that's a two hour meeting and everybody who is available is expected to show up." As this quote indicates, community meetings are scheduled for a regular time period (e.g., monthly) and typically have a structure such as committee reports, actions items for decisions, and community updates. In some communities, these meetings are followed by a meal for additional social time.

Communal meals are a typical event in senior cohousing. Many of the common buildings have elaborate kitchens where members or rotating "cooking teams" take turns preparing food for a shared meal and cleaning up. Both authors were invited to shared meals during the visits. These were lovely times that involved delicious food, bottles of wine, and a great deal of fellowship. Meals included several options that took into account food preferences and needs such as vegan/vegetarian, low salt, and gluten free. It is clear that there is sensitivity to diet restrictions and preferences and a desire to be as inclusive as possible so that everyone can participate. These shared meals are a primary time of interaction for the majority of communities, although the frequency varied. In some communities, there would be a common meal at a set time, such as every Sunday evening. In other communities, these set times would be supplemented with other shared meals, as reported by Chip, "we're a group that does like to get together. We have two or three meals a week opportunities. There's always a Saturday breakfast and there are one or two dinners…."

Although many residents reported that they enjoyed their time cooking together, residents in more than one community decided that they were not interested in preparing large meals for everyone. Tammy reported that an elaborate kitchen was designed in her cohousing community, but no one wanted to take responsibility to cook. These residents found other ways to stay connected and also make sure the kitchen is put to good use. Instead of cooking for the entire group, they have potlucks and rent out the kitchen for cooking classes and nutrition programs which brings in some revenue for maintenance. Likewise, Dixie reported on the decision in her SCC, "Well, people just didn't wanna cook anymore. It's like, been there, done that, too. Didn't wanna cook anymore. And so it has just kind of fizzled out. Right now, the way it is, is we have one meal that we bring in a chef, that we pay for… very economical, 'cause we all have to pay for part of him. It's like $7.00 to $8.00. And then we go out to one of the restaurants in [town], because there's so many great restaurants. And then we started this concept called a pop up dinner where you bring your food, your own set ups and your food from your house, and we just… In other words, you eat with others, but you take care of yourself. Somebody doesn't have to clean up and clean the kitchen and organize."

Besides meals, the SCC have classes and other programs onsite. Several that we visited had special rooms for arts/crafts, woodworking, meditation and devotions, and libraries. Residents with expertise often headed activities for the other members such as leading yoga or forming a book club. Others formed groups around shared life experiences. One of the authors visited a community around Mother's Day, where a resident had established a group for mothers who had lost a child. This shared experience provided a space for these mothers to grieve and support others who understood their loss. In more urban areas, individuals outside of the community were also involved in activities within the SCC such as one where local artists rent space in the common building. As a result, the cohousing community is included in some of the local art walks that happen monthly. Another SCC allows neighborhood groups in which residents are involved to meet in their common room. Through this type of interaction the SCC is becoming more integrated into the neighborhood.

Parties and holiday celebrations also occur. Holiday celebrations provide a time for those who have few family in the area, and also promotes relations within the SCC. The "usual" holidays are celebrated—e.g., Thanksgiving, Christmas, Independence Day—but festivities also occur at other times of the year. For example, a few mentioned solstice gatherings which are part of the community calendar. In addition, parties seem to be organized for many other reasons—or maybe no reason at all. Kim stated, "There's always something going on here. Parties! Sharon and Benjamin have parties for everything. Everything they can think of. There's always a party."

Committee work, in which residents engage to sustain their communities, constitutes another major source of joint activity. The types vary, yet there was a consistent expectation that members would participate in committee work. While chores and tasks are undertaken, committees are another form of social interaction within the community. Jim stated, "[the work teams] are a connection with other people, but then there's also a lot of work involved…..It's all voluntary, but there is some minimum level of expectation. In our group, you have to belong to at least one team and you pick a team that you have some affinity for…. you are expected to show up for the team meetings which are at least once a month." Examples of committees are gardening, social time, devotion/spiritual life, and overseeing the common house and property. The benefits of this social time is summed up by Marge as she states, "one of the things I really love doing is gardening. It is so wonderful in my opinion to be able to garden with other people".

Creative strategies were formed to deal with residents who had to miss committee meetings or team tasks as there is an acknowledgement that people have commitments and engagement outside of the cohousing communities. "..We allow for people to travel. I didn't get my hours last month, because I was gone for three weeks. We allow for that in the overall scheme of tasks." (Andie). This community was thinking of setting up a payment plan option for people to financially contribute if they were gone for extended periods and missed their work details. Andie also gave an example of when people were "excused" from these responsibilities such as a health or family emergency. "We had someone that just died, and his husband, as a task manager, I asked if he could have a waiver."

Spontaneous and unstructured activities. A benefit of living close together is the opportunity for community members to engage in unplanned activities together. A number of people remarked that they would ask others to go into town, see a movie, or do something else together. As one of the community residents remarked, "There's so much going on and people here are interesting people who are interested in a lot of things—opera, going to a concert or the movies. It's not uncommon to see something pop up on our email list that says, "anybody want to do this? …(e.g.) someone snaps up [an extra ticket] and they go to the play together. She got a ride and he got the ticket." A cohousing living environment promotes spontaneous invitations and options. For example, if one person can't accept an invitation there are typically others who could be asked.

As a result of the relationships between the residents, other types of unplanned activities occur (Fig. 5.2). In one community, coffee and conversations happened regularly. If someone was thinking about a particular topic, she or he would send out an email and invite others to have coffee in the common room to discuss the issue. Examples of some of the conversations that took place were political issues and elections, death and dying, and climate change among others. As one resident stated, "…there's a lot of ways to have fun here, and there are also, at different times, conversations. Like somebody will get together. Vera one time had conversations on aging …and then we would discuss them". As this example illustrates, activities within the senior cohousing communities are fun and pleasurable, but also touch on topics that are important to residents who are in the later stages of life.

Safety was an issue that was brought up frequently and the safe environment in the SCCs promoted activity levels. Since others were around, residents felt safe to be out in the communities and did not feel confined to their homes. Mae remarked that "I like being able to walk out of my house even in the black of night, and walk down [the trail] and feel safe." This was important, as she felt secure going to visit her neighbor after dark, which was not the case where she previously lived. Another woman remarked that she had difficulty sleeping and would go to the common house in the middle of the night to do puzzles, and felt very safe. This element of having other people around, who cared about you and noticed your whereabouts, was an important aspect of SCCs for several people.

Fig. 5.2 Happy hour and conversation

5.2 Engagement Outside the Senior Cohousing Community

Residents were also active outside of the SCC, and involvement and engagement with the larger environment was important for many. The founder of one of the communities once did a survey of how residents were involved in the local community and reported that residents participated in 31 agencies within the area. In this way, the local area is a recipient of talent, expertise, and skills of SCC members who participate and volunteer in community life. As Marie states, "This is not like a set-apart community… we've tried to be something other than just a community closed in and upon itself."

In many of the cohousing communities, residents were civically active and involved in local governments or political action (Fig. 5.3). When one of the authors called to set up a visit to the community, the resident who was coordinating the visit stated that he was the only one around that day…. everyone else was at one of the political rallies in town. He jokingly remarked that it might be that a few were arrested before the day was over! A few of those interviewed discussed involvement in political campaigns and many of the cars in the parking lots featured political bumper stickers.

Numerous individuals volunteered in their local communities, and residents often noted their involvement in programs related to social and environmental justice. In one community, the SCC was very involved in Earth Day activities and clean-up in their area. Another community hosted Crime Watch meetings in their common room space. A few people mentioned involvement in community meal programs such as

Fig. 5.3 Participating in social and political issues

home delivered meals or food distribution programs. Others discussed educational programs such as reading and library series, and working in the local prison. Deb stated that she is involved with the fire and rescue league and reading programs among other involvements. "I'm very involved in social justice issues… I've tried to broaden my horizons because I would find that just being here in [my SCC] would not be enough for me." Clearly, involvement in justice programs were a priority for many senior cohousing residents.

Involvement in the arts and cultural life of the local community was also common. As one person stated, "There's so much going on here in the art world. That's another reason that I wanted to move here." Others volunteered in cultural events at a nearby college, such as ushering during films or music programs while others took art classes offered in the local area. Some activities in which residents were engaged were specific to unique challenges that they had experienced. For example, Marge stated that, "I'm doing art work for cancer survivors and a writing group." As can be seen, cultural opportunities located beyond their own community were important for many of the members who were interviewed.

While some of the SCCs have strong spiritual emphasis, many residents were also involved with spiritual and religious activities beyond their cohousing community. One of the women was a part-time pastor, others attended churches and were members of religious communities, and some participated in religious-related programs such as thrift stores and food pantries. Several of those interviewed attended meditation and yoga classes in their area as well.

A move to a SCC can create changes in social networks. One example is trying to retain friendships with those who live outside the cohousing community. As new friendships develop within the SCC, changes may occur in existing relationships. Betty described this tension, as her birthday was approaching, "I don't see my quote "old friends" as much as I used to. So it's changing the social system because I do hang out with folk. I'm going to have a birthday next week and I'm looking forward to at least two or three celebrations.... I think that I'm not unusual that I've kept old friends and I've lost a couple people in doing that....That's how life is.".

5.3 Public-Private Spaces

As discussed, it is clear that residents engage in a variety of activities within and beyond the cohousing communities. However, those interviewed also discussed their need to structure private or alone time. Although residents report high activity levels, many stressed that they valued and protected their "me time" and space. Several commented that the majority of those living in these communities are introverts, which might seem counterintuitive for a communal living arrangement. Kim explained this situation very clearly, "for introverts it's perfect, because you go in your house and you can be in there as much as you want, but when you come out, you don't even have to go make friends somewhere. You're just here with whoever you want to be with that's here." Introverts require space and time to refuel and refresh, and need to have quiet time away from others. Those interviewed reported several different strategies for claiming alone time within their cohousing community.

In some communities, there were clear boundaries that created quiet spaces. For example, one resident discussed how front porches of homes were public spaces—if people were out, it was appropriate to say hello and chat. However, back porches were private spaces. If you saw someone on the back porches, you didn't engage her or him in conversation. "If they wanted to talk, they would be on the front porch. The back porch is the quiet space." Likewise, those living in a condo also had similar norms where residents would open their front shades if they were available for company. Closed shades meant that it was private time.

Norms and practices also dictated how private space is respected within the communities. For example, neighbors do not simply walk into each other's homes. Even those communities where people rent out part of their home or had an apartment in the common building, there are boundaries and demarcations for the private areas. Additionally, a few people mentioned that if they needed their alone time, they would simply ignore a knock on the door if someone came by. It seems that this practice does not create hurt feelings or bad will, but is a practice that is respected as a "private time" strategy by those in the community.

For couples, there is the added dimension of accommodating different personality types and need for social/alone time. Multiple couples commented that one person is outgoing and the other is more private. Larry gives a great example of how he and his more extroverted wife experience interactions differently within the cohousing community, "Jenny had been down to the office for something and then she told me, she came up with characteristic enthusiasm saying "Boy, all I was doing was walking from the office to the elevator and I ran into three people that we got involved in conversations and it took me a half hour to get to the elevator. Isn't it wonderful? And I'm thinking, oh my god. That would be torture for me. I'm trying to get to the elevator, and yes, I would stop and talk with those people. I probably wouldn't have as long a conversation and so on and so forth, but to be polite, I'm not going to brush them off…"

In summary, the residents described their communities as places where a significant amount of formal and informal activity takes place. This was one of the pull factors to move into a cohousing community. There were numerous planned activities to bring people together, for fun, for social time such as common meals, and for maintaining common spaces. In addition, however, spontaneous activities and events were frequent occurrences as the environment promotes interactions among the residents. While activity levels within the cohousing communities were high, the surrounding areas also benefited from having a senior cohousing community their area. Residents were energetically engaged with programs and events in the larger area where they contributed their time and talents. However, the need for private time and space was also well recognized and respected. Residents had strategies to recharge within the context of their shared community lives. Overall, residents reported a good balance of together and alone time which provide engagement without being overwhelmed.

Chapter 6
Benefits and Challenges

> [The best part of living in senior cohousing is] the connections as I choose to have them or not. I can initiate contact if needed, I can participate if somebody needs me. I think just the accessibility, the fact that we're accessible to each other, for each other. (Edna)

> The differences of opinions haven't gone away, but one of the pluses of co-housing is that you have intention to get along with each other and ways of working out conflicts. (Mae)

Residents of senior cohousing communities readily speak of the benefits they've experienced while living in a SCC. However, they also openly acknowledge the challenges that they and their particular communities have confronted. When asked in what ways, if any, they have grown as a result of living in senior cohousing, the residents' responses reflect their experiences of embracing both the advantages and the struggles that are inherent in this particular type of housing venture.

6.1 Benefits

The senior cohousers were enthusiastic when speaking of the benefits they had experienced living in their communities. They shared many stories of their involvement that shed light on the strengths of living in a SCC. The positive aspects they discussed were both social and practical in nature.

S. Cummings and N. P. Kropf, *Senior Cohousing*, SpringerBriefs in Aging,
https://doi.org/10.1007/978-3-030-25362-2_6

6.2 Social

Not surprisingly, the major advantage of living in senior cohousing cited was social. After experiencing shrinking social networks while living in their previous homes, the ready availability of engagement and interaction was a highlight for many. Marie stated, "...the benefit is having an immediate community. You have people around." while Mae reported "...the ease of sociability is one of the big ones for me. Anytime I wish to have social interactions, it's easy to find them, and to know all of my neighbors well and to feel like I can call on them. I know their names."

In cohousing, socializing is easier as the proximity of others promotes interactions. Previously friends might have lived 20 or 30 min away and in order to see them it was necessary to schedule a time to get together and then drive a distance to meet them. Many people noted how much they appreciated the fact that when they wanted to socialize or felt somewhat lonely, they could simply walkout of their apartment, run into someone and have an interesting conversation. Others commented on the ease of socializing with others who live so close by. There is no need to drive or try to find parking when friends live a few doors away. "Well, the benefits are a ready social life. I mean, we laugh, and we have dinner parties, because we walk 50 steps." (Andie).

Spontaneity was also seen as a benefit by many. If someone had an extra ticket to the theater or suddenly decided that they'd wanted to go see a movie, for instance, they could send an email through the community's listserv and find someone who would like to join them. Pop-up dinners and outings also naturally occurred. Jenny noted that she and her husband frequently invited others over for dinner, "There's so many people to invite, which I do constantly, spur of the moment." "What are you doing? Want to come for dinner and to share the results of my experiments at potluck?" Linda also commented, "Yeah, I agree. I think it's just nice to know that there are people around. Like tonight my husband is having dinner with a friend, so I called a friend here and we're going to get together. So, it's just nice that I have people in my own community that I can spend an evening or an afternoon with." (Fig. 6.1).

While all types of social interactions may be important, many of the residents clearly treasured the opportunity to easily develop and maintain close friendships. The proximity to others who have shared values and interests fosters this ability. "For me the benefits were meeting new people who are quite intelligent. That was a real plus. Getting to know my neighbors quite well as compared to where I used to live and just occasionally saying hello. It's much more in-depth here." (Chuck). Many spoke of the importance of emotionally connecting with others on a daily basis "...being in close proximity with the people that I'm emotionally closest to is really something I appreciate a lot. If we were scattered around and then we just gathered together for meetings or something it wouldn't be like seeing the people every day." (Faith).

The easy availability of social interactions and activities can also benefit married couples. The media is replete with stories about the stresses that can occur in a marriage when both individuals are retired and living together with little social engagement outside of the home, (e.g. Akitunde, 2017; Goodman, 2011). One resi-

Fig. 6.1 One of the married couples in a SCC

dent noted the positive impact that living in senior cohousing had on her marriage, "My husband and I have outlets, so we don't have to get on top of each other because there's nobody else to vent to… I think our marriage is stronger and it wasn't really in trouble at all. But there are so many more places for each of us to go. It's not just two people stuck in the little living room…there's more room for members of a couple to do different things and talk to different people and have different conversations." (Jenny).

This sense of social connectedness which is the heart of senior cohousing communities, benefits residents on several different levels. One of the men that we interviewed stated "…there are three diseases of life for seniors. One is isolation. The other one is boredom, and the third, which is really a sneaker, is uselessness." (Pete). As can be seen, one of the major goals of senior cohousing is to promote social relationships and combat isolation. This goal is clearly being met for many residents. Being with others who are active, participating in activities within the SCC, and having access to events and opportunities in the outside community also helps keep residents intellectually and physically engaged. Lastly, connectedness is a remedy for feelings of uselessness. Residents reported that it was important to not only know that others were there to help them; but, so was being there to help others. As one resident stated "I think one of the biggest benefits is being able to make good friendships with people that are right close to you. …when you need somebody, somebody's there for you. Or if somebody needs you, it makes you feel good to be there for them.". Samantha concurred, "Meaning, you know, when you care for others, that gives you a tremendous amount of meaning. And there's a lot of people to care for here and to care about. You know what I mean? It's like family."

6.3 Practical

The availability of practical support and assistance was another benefit for which residents cited strong appreciation. Knowing that others were there and ready to assist with everything from small needs to major emergencies was deeply meaningful to the senior cohousers. Being able to "run right next door" to borrow a cup of sugar, some aspirin or even a bottle of wine was commonly mentioned. One resident recounted the time when she ran to her patio after cutting her hand and shouted "help, does anyone have a band-aide" and immediately several of her neighbors were there, band-aides in hand. Others who previously had sole responsibility for the upkeep

Fig. 6.2 Working in the common area

and care of their own homes and yards, expressed relief at being able to share this obligation. "I think I feel a lot more relaxed about everything because I don't feel that I have to deal with everything by myself." Senior cohousing members are active individuals who, in general, wish to share, not relinquish, all responsibility for the care of their environment. Residents combine their knowledge and skills to maintain the common areas and their homes (Fig. 6.2). As one resident noted "…there are so many skills and experiences here that if you need to learn about something or how to do something somebody's going to know and they're going to help you and they'll probably just do it with you. So, a huge benefit for me of a senior community is all of these lifetimes of experience in one place."

Another practical benefit that was often discussed was residents' willingness to help one another out with small services such as providing a ride to the airport or the doctors, pet sitting, picking up packages and going to the grocery store. "People give rides to the airport, people dog sit… take the dogs for walks. If I know one of my neighbors has difficulties getting out for whatever reason, I'll call her and I'll say, "Hey, I'm going to go to Trader Joe's, do you need anything?" You know, vice versa, someone will call me…" (Ava). Andie also reported ""Pick up and drop off. I mean, I was going to take Lyft home from the airport yesterday, and that was like $50. Oh, no, I'm just going to do BART; but, I had a big heavy suit case. And, so, while I was on BART, I texted two friends. I said, "Anybody available to come pick me up from BART in a little bit?" And Cara said, "I'm actually out in my car. I'll just meet you there."""". This type of practical assistance greatly enhances ease of daily life for the residents and also provides a sense of interconnectedness. When there are others available who care enough to help you out and who you, in turn, can assist, you know you're not alone. "…that's a real strength of this community is the being there for others when it's needed, yeah." (Linda)

6.4 Challenges

While the participants we interviewed were mainly very positive about their experiences living in senior cohousing, they also readily discussed some of the challenges they face. These challenges fell into three categories—interpersonal relations, time requirements and self-governance.

6.5 Interpersonal

While living and engaging with other people is one of the biggest benefits of life in cohousing, it can also be one of the greatest challenges. Not all people are going to have similar temperaments or personalities, and some may clash. "And, it's not all… You know?… It's a mixed bag. It's not totally harmonious. Some of us get along with each other, better than other people do." (Mae). This is a reality of group living.

A number of respondents noted the personal challenge of tolerating those who are different and who they may not like, and not assuming that others should change. "The challenge to me is, it's acceptance. And to me, what that means is to be able to accept other people's foibles and their incapacities." (Jewel). Ava expressed a similar sentiment, "Yeah, and I think accepting the fact that people are different, in accepting our differences, rather than trying to mold everyone into one type of person".

However, conflicts between individuals do arise. In those cases, not only the individuals themselves but the entire community may be impacted. How communities deal with such situations vary. Some communities reported having retreats, workshops on effective communication or bringing in an outside mediator to help resolve difficult interpersonal conflicts. Other communities, however, are not as ready to face such conflicts directly. In these cases, interpersonal difficulties may be glossed over. "We haven't worked out the problems that we have with one another. The problems with the building are one thing… But the interpersonal difficulties that develop we don't deal with." (Monica). While we didn't see any evidence to indicate that the functioning of these communities was threatened due to this dynamic, it did appear that the group cohesion in cohousing communities that did not deal with conflict in some way was less strong.

Differences of opinions and conflicts can arise when important decisions are being discussed and decisions are being made. Some people are more adept at interpersonal communication while others can be abrupt or rude. As a result, feelings may be hurt. One woman explained, "…and we're all very, very intelligent people, experienced in so many different ways, and oh my goodness, so many smart, knowledgeable, experienced people here, so we have meetings and some of us are more opinionated, and just, you know, "Well yeah, it's gotta be this way, because my experience says you do it this way," and we've experienced that in more than one team, lots of teams." (Ava).

One challenge is to maintain a commitment to the community even in face of conflict and difficult times. While some people do withdraw, a community is maintained and prospers because the majority are willing to devote the emotional and practical time to work though conflicts. When speaking of community meetings, Mae noted, "Other times, it's been …contentious, and it has been a challenge at times, and then you don't feel very good after you've left the business meeting. But, that's just the dance of human relationships, and we're still in there dancing with each other to maintain this common property."

6.6 Amount of Time Required

All residents readily acknowledged the large amount of time taken up with meetings when living in cohousing. Because residents jointly make decisions and are responsible for carrying out the work of the community, they are asked to attend both monthly community-wide business meetings and smaller team meetings. Although this is true at any time point in a community's life, it is especially true in the beginning when a smaller group of members is working to establish the community's mission, structure and processes. According to some who were involved in the development of new communities, spending so much time in community-related activities encroached on their other activities and on individual preferences such as traveling. "There were so many meetings. It really was eating up a lot of time… we were so involved in the community because we wanted to set up these systems and we wanted to go these meetings and there were so many meetings… Our second year was in March, and so many of our systems that we worked so hard to set up have been established, so now the meetings are lessening… (Ava).

Some participants reported that their main method of interacting with others was via work and that they really enjoyed team meetings where they could engage with others to get things accomplished. However, not everyone was so positive. "I think one of my disappointments, in a way, is the fact that it's so time consuming. I find that I spend a tremendous amount of time on different committees and so on, and sometimes I get resentful about that." (Deb). As mentioned in Chap. 2, variations in amount of time that members actually devote to the community exists. Lack of participation, however, can also cause resentment. "I've seen the result of some people that don't like meetings and they come and they get up 45 minutes into the meeting and saying, "I have to leave. Got some place else to go." It's not necessarily that. It's that they confess later, "I hate meetings."" (Mary)

6.7 Self-governance

Also as discussed in Chap. 2, decision making in co-housing is consensus based. While many see this as an advantage, (i.e. everyone having a voice), most will also acknowledge the difficulties that this entails. First of all, people note that this type of process can be a very slow one. One person compared it to "watching paint dry". Another stated "One of the challenges to me is the pace at which things get done. I consider myself an action cohouser and can't get used to the amount to time that goes into getting any decision made. It's slow…and one of the things I'm not accustomed to—is that, is the lack leadership that exist in this community by the nature of having to go with consensus. You can't take the lead in many things without being yelled at for being an authoritarian personality; so, you have to be much more considerate when you want to advocate for something. You can't just assume that because it seems like a good idea to you that it will seems that way to other people." (Patty). As is obvious from this quote, for those who are used to moving quickly or are accustomed to making decisions and having them followed, it can be quite a challenge to take things at slower pace and allow everyone to have a voice. This same challenge can be felt by those who were used to having the home domain completely under their control, "I think it's difficult for some Americans to understand the idea of consensus and learn how to balance life together. I know all the 50 years I was married and raising the kids, the house was my domain, and I decided what we are going to eat, and I decided when we were going to go places and, you know, how the house was going to be kept, how the garden was going to be kept. All of that was my decision and nobody challenged but here everybody's got an opinion." (Regina).

As previously noted, not all communities define 'consensus decision-making" in the same way. But, the process of defining what is meant by consensus can be difficult, "Consensus is assumed to be of the form that many Quaker communities have but, the ideologically strong part of this population [i.e. community] wants unanimity and they do not want to have majority rule. So, they're struggling to see if they can define a form of consensus which is not majoritarian, and which is nevertheless on some level unanimous. They want to find a way of defining unanimity so that it doesn't coerce agreement." When consensus decision-making is defined as 100% agreement, difficulties can arise because even one person disagreeing can block a decision from moving forward. Larry reported "But working within a consensus, you can't vote the other way and allow the group to go on at all…And so you either have to subordinate your own voice and say no, I'm going to go along with that, or you have to mess up the whole enterprise, which you're probably not willing to do. So, most of the time, you end up much more subordinating yourself in that kind of a system."

Communities handle dissenting opinions and votes in different ways. In one community residents mentioned, within their culture is the expectation, that a negative vote must be based upon what one believes is in the best interest of the community and not on personal preference alone. In another community, when an individual or small group disagrees with a proposed decision and agreement cannot be found during the community meeting, "… a facilitator and interested parties will discuss it with you and see under what circumstances you would remove your block … Then if after three times, if it hasn't been resolved, we go to a super majority, which is 80%." (Trent). In another community, when there is disagreement, those interested in the issue will gather together to see if alternate solutions can be found or a compromise achieved. If so, they then bring the proposed solution back to the larger group.

Some of the most contentious problems seem to arise when discussions spontaneously arise during a community-wide meeting or are presented with little preparation. In order to avoid these situations, some communities have worked out processes for bringing proposals to the community meeting. In some SCCs, a small committee is formed to run the monthly meetings and those interested in advancing a proposal first bring their ideas to the committee for feedback, thoughts on how to improve the proposal or suggestions for research to undergird the proposal before it comes to the floor. In other communities, those interested in bringing up an idea at a community-wide meeting will first survey community residents to see what reactions they may receive. Questions, negative comments and suggestions are all used to refine an idea and at times, members may engage in preparatory education before an idea is brought up at a community meeting.

In spite of this, all members of a community may not agree on a certain decision. As one member indicated the real challenge is for anyone who grew up in the U.S. to move from an individualistic "position-based mode" to a "concern-based mode" (Bob). In the end, as Bob suggested, the best question for cohousers to ask themselves when faced with a choice they're not fond of, is not "Is this what I really want" but rather, "Can I live with it". If the answer is "yes", then he suggests that the member do what is best for the community and let the decision move forward.

As can be seen, life in senior cohousing communities involves both great benefits and some real challenges. The joy of community is living close to others for whom you care and who care for you. The flip side of this benefit is the proximity of and the need to work with others who you may find irritating. Likewise, the ability to participate in a community that fosters independence and active engagement among like-minded individuals can be a true gift. This ability, however, is also accompanied by the need, at times, to let go of individual desires and to work for the good of the whole.

References

Akitunde, A. (2017). Retirement can hurt your marriage (And What You Can Do About It). *Huffington Post*. https://www.huffingtonpost.com/2013/08/09/divorce-after-50-retirement_n_3286342.html.

Goodman, M. (2011). *Too much togetherness: Surviving retirement as a couple*. Springville, UT: Bonneville Books.

Chapter 7
Looking to the Future—Aging in Place

> My major thing is that as the years go by and one gets older, one starts thinking more and more about the need to be in a support system, and I want to be a support to other people. I'd like to have a support system around me, and I don't want to run out of choices…. (Mae)

While previous chapters have summarized why people move to senior cohousing and their subsequent experiences, this chapter will examine how senior cohousing residents see their future as they age within the community. An important issue brought up in numerous conversations was ability to receive and give support. This chapter will summarize thoughts and hopes of residents as they look into coming years.

7.1 Hopes for Aging Within Senior Cohousing

As Mae identifies in the quote above, senior cohousing provides older adults with a support system that surrounds them. Several people discussed the desire to be among others, and to not be by themselves as they age and face the end of their lives (Fig. 7.1). Noreen states, "There's a sense of ease knowing that we're not gonna be alone. Knowing that there are people around who care……. That's the way I wanna live and the way I wanna die." Stephanie echoed this sentiment with a desire to end her life within a support system where she has held membership and been cared about. She stated, "I want somebody to know me when I die. I don't want to die alone."

S. Cummings and N. P. Kropf, *Senior Cohousing*, SpringerBriefs in Aging,
https://doi.org/10.1007/978-3-030-25362-2_7

Fig. 7.1 Connection and care

Part of the hope of living out one's life in a SCC is the opportunity to age with dignity. This idea includes being surrounded by individuals who accept you as an aging person. Marge states that "I wanted to be able to age gracefully and stay in one place doing it. Both the social aspects of having people around and having a nice place. Being able to [age] without having to be in a nursing home or care facility that I've seen other people be in." This sentiment represents a way of thinking about long-term care (LTC) as something to be avoided—the feeling that LTC strips an individual of her or his dignity and is more of an institutional, versus home environment. Others also indicated that they could not envision themselves living in a long-term care setting.

The cohousing communities provided space for the residents to discuss aging which promotes clarity and decision making as they look forward. The process of being with others and entertaining these important, but often unspoken issues, is valued (Fig. 7.2). Lucy stated that "we've had so many useful meetings or opportunities to talk about aging, and what it has meant for us, and what we expect it will mean for us in the future, and how to be ready for it. That's one of the most valuable things I've found in senior cohousing."

Fig. 7.2 Pets are part of the family in SCCs

7.2 Support in Senior Cohousing

As people discussed growing older within the cohousing communities, they stressed the importance of giving and receiving support. There are various reasons that people receive care as they age within the community. In addition, the cohousing communities had different types of structures to address support needs—ranging from ad hoc and informal processes to more elaborate and structured approaches.

When a community member faced a health challenge, others often became involved in multiple ways. Beneath these practices is a social norm that when an individual requires assistance, others in the community will provide aid. There is an expectation that people will be supportive and in return, there is security that help will be available when they need assistance. Mary discusses the reciprocal belief about care when she was in a wheelchair for six weeks, "Everybody here is very aware that they want to help me because they need help every now and then and they want the support and they want the willingness and compassion of others to be there and available. You just understand that that is what we're about there."

Those interviewed shared several examples of how support was provided within their cohousing communities. One woman recounted how the community became involved immediately when she was taken to the hospital. "Bridget followed the ambulance all the way down to the hospital and stayed with me until my daughter came [from out of town]. She came right with me and stayed in the hospital. It was so wonderful".

Support is also provided when people are convalescing from a health challenge. While a person may be well enough to be back home, there still may be functional limitations in what she or he is able to do. If a partner/spouse is available and serving as the caregiver, the community rallies around the couple by delivering meals, running errands, and providing social contact. Billy recounted a story of a couple that needed different furniture after someone had shoulder surgery. "So, people said, 'Oh, I've got a recliner you can use to sleep in; let's move it to your house.' That kind of thing is just matter of fact."

Support is even more crucial when an individual lives alone and there is not another individual to assume the primary caregiver role. This was the case with Pete who had no family in the area. After suffering a stroke, "I was rushed to the hospital, and then in rehab for a couple of weeks. I didn't do a thing. Others came in my house, they cleaned out the fridge. They arranged for a companion [upon discharge]. All I had to do was worry about getting well." Even relative newcomers to the community receive extensive care if they require it. One person described how extraordinary the experience was for her. "I had surgery after I was here for a year. It was just totally not within my realm of knowledge or experience that people that I considered neighbors came to help. I had somebody doing the laundry and somebody help me with food, and somebody changing my bed. I never experienced anything like that and I didn't plan on it. I was prepared to hire somebody to do this for me but I didn't have to. It was just an amazing experience."

As part of the discussions about providing assistance, there was also a sense that individuals were "duty bound" to take good care of themselves so they would not overtax support systems. Francis described drinking green smoothies and taking vitamins to keep his health as good as possible to prevent others having to become his care provider. Additionally, one of the men described a health crisis and the resulting change in his behavior, "I was in the ICU for a while and then rehab. One of the things that this has done is made me realize that I have a strong obligation to the community, to be well, not put everyone through all the grief again. Now I go to the gym five days a week and do yoga." An aspect of the norm of support is staying as healthy as possible so others will not have to be in an unnecessary caregiving situation.

In addition to health-related experiences, older adults face other challenges that require support and assistance. One area is loss, such as becoming widowed. When a spouse or partner dies, a primary need is to emotional support to get through the difficult time. Samantha describes the loss of her husband, "When Doug died….. other people saw what was happening and within minutes they were there. You know, even before the end they came. Sitting around the table and giving me support. I mean – where do you get that kind of community?"

Another loss that is experienced is the onset of dementia in one of the residents. This situation can be complex as the disease impacts functioning in multiple ways. However, people with dementia, in most instances, were living with a spouse as those who were living alone with this condition often left the SCC for a more supportive environment. For those with a spouse or partner, other residents were willing to assist with tasks associated with care provision such as providing respite, companionship, and running errands. One person described how others recognized the stress experienced in providing care… "then sometimes people decided that I needed a mental health day, so they took me out to lunch."

The various cohousing communities had different models for providing support. In some SCCs, the community expectation of support and aid for each other is an informal process. While the principle of support exists, there is no committee or formal process about how assistance is provided. Dottie described this process in her cohousing community, "We as a group haven't set any policies, it's more individual. People here are very helpful when someone has a medical emergency and taking people to the doctor or getting food for them. That kind of thing. But there's been no formal acknowledgement about what people would do."

Other senior cohousing communities have formalized care structures as part of their organizational plan. Pete indicated that his cohousing community has established an extensive resource list of services in the area that residents can contact, as needed. However, he also serves on a resource team, which "establishes a network of help for people who need it within the community…. We have to put our money where our mouth as… we're dedicated to having people age in place. You can't not work on that". Stephanie describes a complex care structure which begins with every resident having a personal file with a living will, medical chart and other relevant information. In addition, this community has a chart with activities that are provided to members when there is a health situation, such as doing light housework, transportation, taking care of a pet. Other community members sign up to take on these tasks. In the formal support model, the provision of assistance is situated in a committee or group which takes primary responsibility to make assure that help is provided to the individual who needs it.

7.3 Limits of Support

As described, there is a great deal of support provided within senior cohousing communities. The goal is to allow individuals to stay in their homes and SCCs as long as possible. However, there are limits to the amount and type of support that is expected in these communities. It was clear from the community residents that extended or complex caregiving was outside the bounds of what is expected of others. A number of people commented on this issue. Kim stated, "We do have some intention of helping each other through temporary conditions, but … if it's going on for years or sort of indefinitely, there would be limits to what we can do." Another resident stated, "We're not set up to be assisted living. It's really easy to fix a meal and take it over, and do laundry or drive somebody to the grocery store, but if you needed full-time nursing you will have to hire someone."

One of the limits appears to be when someone requires intimate caregiving. Pete stated that, “Martha broke her hip. We cooked for her. But we won’t touch you. Nobody’s going to come in and bathe you. That connection has to be professional.” Deb described her responsibility to care for her 100 year old mother who lived with her. “I can take care of her, and I don’t expect the community to. I wouldn’t expect my neighbors to do that. I mean, that’s just asking too much.” Betty summed up the limits of care succinctly, “Aging in place is what we do – everything up until personal care.”

In order to handle the care responsibilities in some the communities, people were hired to assist in various ways. In a couple of the SCCs, maintenance of the community was beyond the capacity, or interest, of the residents and certain chores were hired out. Sometimes this exchange involved a type of bartering relationship—the community would provide someone with something in exchange for doing work. Tammy described the situation in her SCC. “We have hired a handy man and we have a deal worked out with him where he can use our shop area when he works here and in turn he gives us a break on what he charges. Instead of us doing all the repair work, you know.” Faith described how the community used an unoccupied apartment to bring in help. “We have a little apartment that’s upstairs in the community house….. It’s a couple with a preschool girl. So she [the wife] does some things around here for us.”

For the most part, having residents who are in on-going poor states of health is somewhat unusual. When someone does require assistance, however, help can be brought in from outside the community so individuals can remain in their homes. Lily described how one of the residents in her community was able to stay there even in a state of very poor health. “Sue is in a wheelchair and she has some weird neurological thing and she can’t use her legs at all. Now she has 24 hour care there in case she falls. …. So people have assistance. One lady even has hospice help. That’s all from the outside, you know.” In another community, the average age was 90 years and collectively, the residents hired a physical therapy assistant. She worked with several residents on the given day she came to the community—which was convenient for everyone (Fig. 7.3). Not all residents have the financial means to pay for on-going in-home help, however. In cases in which a person has continuing personal care needs or progressive memory loss and cannot afford to hire permanent help, then a move to a more supportive environment will eventually become necessary. In several of the communities, previous residents had made the decision for financial or other reasons to move to an assistant living community. Members of one community noted that two former residents had relocated to a nearby assisted living facility and that current SCC residents who knew the two would go to visit their former community members.

Fig. 7.3 A physical therapy session

7.4 Envisioning the Future

Thinking about their future, the residents were clear that part of the appeal of senior cohousing is that they are in a collective of aging individuals. In this way, there is security about the future. Ava stated that "at least one person, if not more, will be looking out for us, as we will be doing with those around us. As we grow older, we will be ok. We are in good hands."

Future considerations include more than simply attending to the physical changes that accompany aging. John states that, "The support extends beyond providing physical health to people when they need it. There is a kind of energy around people that are here already. … It's important because everybody knows if they really need it, they can tap into it." Suzanne also elaborated on that point, "We know what's going to be here for us. We don't have too many elders that require care yet. But it makes us feel more secure as we go into that phase."

While there is an acceptance of aging, and a security in living among others in the same life phase, there are also questions about how to meet some of the challenges. Nancy described the aging of her SCC. "I've been on the dying at home committee… and know we're getting older and physically can't do what we use to do….. so, we don't know [what the future will hold]". In particular, those interviewed described some of the struggles of having residents with multiple and complex conditions. "We have one person who's dealing with an encroaching dementia and her loss of her short term memory. There are difficulties in dealing with that. And that will get worse and it's going to incur in more people. I don't exactly know how we're going to deal with all of this. We're going to have to see as it comes." (Larry). The concern about potential memory loss among community residents was brought up several times. Because the SCC movement is relatively new, not many communities have had to face this issue. While expressing concern over this possibility, community members, on the whole, were hopeful that just as they had dealt with so many other dilemmas in the development and running of their SCC, they would be able to create feasible and effective approaches for facing this challenge as well.

The reality of living is that at some point, the future will include the end of life. Within the interviews, it was clear that death and dying is discussed freely within senior cohousing communities. A few examples demonstrate the comfort that people feel in talking about these topics. At one of the cohousing communities, the residents had prepared for the death of a woman with dementia, Joy, who had been hospitalized and was not expected to live. In the tour of the community, the art room contained a cardboard casket covered with beautiful magic marker drawings. The woman offering the tour explained that community residents had decorated the casket to bring Joy back home to say good bye to her. Remarkably, Joy recovered and was back home—and still quite alive! As a result, her husband suggested that they keep the casket "as she is going to die sometime". Instead of being a symbol of despair, the beautiful casket represented the love and care that the community felt for Joy and was not viewed as macabre reminder of death.

A second example demonstrates the sense of agency and empowerment one of the residents had when facing his own death from cancer. Darren decided that he would have a celebration of his life when he was still alive and had a gathering within his cohousing community. After that, he took his own life on a pre-determined day. In a group interview, several individuals chimed in to tell the story—it was a very positive and moving experience for them. "We had a roast for Darren and we got to sign a coffin top that he had brought down and put in the multi-purpose room. He built his own coffin….a couple of days later, he took the compound of drugs. He told us and his family too. He made sure that everyone was with him on this… He was a man of great self-discipline and he handled himself beautifully for the whole journey. It's something to look up to. The way he exited life was so positive. It was life affirming actually."

Overall, the majority of those interviewed remained optimistic about the future and aging within a cohousing community. There was a sense that they would be able to meet the challenges of the coming years and a steadfast desire to remain in the SCC as long as possible. Pete summed it up this way, "… this [senior cohousing] is a model for seniors to live independently and be recognizing that they're going to get frail. Then we can prepare for it in a lot of different ways."

Chapter 8
Advice from Senior Cohousers

> You're joining a social group, not just buying an apartment. (Tom)
>
> Leave your ego at the door. (Frank)
>
> It takes courage to be happy. (Ava)

The purpose of our visiting senior cohousing communities and interviewing residents was to learn about their authentic lived experiences. After listening to residents' stories concerning what they loved and struggled with in their communities, we wanted to hear what recommendations they would offer to others who were considering a move to senior cohousing. Residents eagerly provided heart-felt advice, born of their desire to help guide other older adults who are looking for community and, also, to help ensure that potential new residents are those who can contribute to the life and well-being of a community. The advice given falls into three board categories—do adequate research, engage in honest self-reflection, and be courageous.

8.1 Adequate Research

The most common piece of advice was to visit multiple communities, spend time and get to know those living in each. Chuck suggested, "If you have a place in mind. Find out as much as you can about it and just talk to a lot of people, like, "What's it like to live here? How much work is there? Do people get along?" All that." Communities differ in how they deal with inquiries from potential new members. Larger SCCs often have membership committees that will communicate and meet with those who are interested in finding out more about their particular community. Smaller communities may offer the opportunity to speak with a designated contact person. All will provide tours. A larger community that one of the authors visited had a specific day of the month when they offered an "orientation" to senior cohousing and provided a tour of their condo facility. The purpose of the orientation was to educate

S. Cummings and N. P. Kropf, *Senior Cohousing*, SpringerBriefs in Aging,
https://doi.org/10.1007/978-3-030-25362-2_8

visitors about the nature of life in senior cohousing and to give them an opportunity to have specific questions answered. Visitors were then invited to participate in a community dinner after the tour so that they could meet a variety of residents and get a "feel" for the personality of the community. While smaller communities may not have formal membership committees, they also encourage those who are interested to come for a visit. "We really insist that people come and look and get acquainted and meet people, and I think that is really important, and talk to more than just one person. Talk to different people and get your questions answered." (Eve) (Fig. 8.1).

Once inquirers have narrowed down their search to a particular community, it was recommended that they get to know the residents who live there "...get to know it before you make a decision. And, really try to get to know the particular people and the particular circumstances. 'Cause they're all very different, ... they'll all have

Fig. 8.1 Open House at Wolf Creek Lodge

different flavors." (Billy). Many suggested that those who are interested investigate the possibility of visiting the community for a longer period of time. SCCs often have a guest room where potential new members can stay for a day or two. Some participants suggested that those interested stay for longer periods of time when possible, by renting a unit for a couple of weeks or engaging in a house swap. If spending a more prolonged period of time is not possible, it was recommended that those interested make repeated visits to find out more about the community and residents. Advice included going to the monthly business meeting, learning about the community's decision-making process, participating in a work day, attending dinners and just "hanging out" with residents to find out about their interests, activities and values.

Spending time at a community may lead to the decision that this type of living situation, or this particular community, is not for you. "…we've had people stay for a week or two, and we both parted saying, well, you decided this kind of living isn't for you, and we agree with you. We got to know you well enough to see that you're set in certain ways of what you're looking for, and it's not here, and it took that long for you to decide. You couldn't decide it in a day or two. It might've taken a bunch of days of visiting." (Francis). Alternately, of course, a longer visit may confirm your belief in your goodness-of-fit with the community and surrounding environment. "I tell them that they buy the community, not the home. That's the exact opposite of middle America which buys the home but not the community." (Pete).

8.2 Self-reflection

Living in senior cohousing is an advantage for many but it is not for everyone. Residents encouraged those who may be considering senior cohousing to honestly consider whether they have the temperament to do so and the desire to fully participate. One resident suggested, "Inventory your needs very specifically in terms of what you can give and what you need to receive. Be very honest. Take an inventory of yourself, and be very honest and authentic about what you have to offer and what you need to receive." (Edna). It is important to know if you're the type of person who prefers participating in outings that are planned and organized by others and who appreciates having a plethora of regularly scheduled activities in which you can participate but for which you have no responsibility. If so, senior cohousing is probability not for you. One participant, Betty, recalled speaking with a woman who was very "gung-ho" about moving into the SCC but who was bothered by the work requirement. When exploring this further with the woman, Betty noted "She said, she has friends down the way about ten minutes and they pay more money than here per month, but they have an art director, they have tours every week, … but all of that they have to pay people to do that for you and they plan it." And I said, "Well I guess you'll have to go to the [Life Plan Community]. And she did." Jenny's advice about areas to consider for self-reflection was very sage "…Ask yourself if this fits your own needs. At the same time, take the opportunity to look at yourself. Have you had issues with getting along with people? This should not turn you away, if so, but see such a community as a place to learn and grow. Ask seriously how you can contribute, are you a giver or a taker? How much privacy do you require? Can you say no or are you strongly influenced by others' viewpoints? An intentional community requires a lot of generosity and time, forbearance, consideration, even forgiveness and letting go. Can you … reserve judgment, adjust in a situation easily, be accepting? Are you able to ask for help when needed?"

Flexibility and tolerance. One of the most common pieces of advice given was that those contemplating cohousing be flexible and tolerant. Residents often mentioned the need to be open to others, to be respectful of those who's personalities are different, and to be accepting of the unexpected. Jewel advised, "There has to be a certain level of cooperation, there has to be respect for each other's privacy. We have to be mindful of what we're saying and listen to what they're saying and receive it.". Others highlighted flexibility and tolerance in relationship to the community's consensus decision-making process. While this type of self-governance allows for maximum input of all and often contributes to creative solutions, as mentioned previously, it is also time-consuming and, at times, hard work. The ability to effectively participate in this process is necessary both to avoid personal frustration and to effectively contribute to the community. Noreen summed up of the feelings of many when she stated "Be prepared to be flexible. Do assume good intent. Another motto, "I can live with it." Not, it's exactly my perfect thing but you have to be willing to, it's give and take, like in a marriage …And you have to be okay with that.". Mary recounted the story of one resident who eventually made to the decision to leave, "I think he realized that he was not prepared for consensus decision making. That's where it really hit him hard. He went and learned a lot about consensus decision-making, but it didn't help because emotionally in his thinking, he was still back on the Libertarian wavelength."

Realistic expectations. Others stressed how important it is for those considering senior cohousing to have realistic expectations and take the time to examine their assumptions and beliefs before making a final decision. Disappointment or frustration may occur due to unrealistic concepts of community. Some may expect higher levels of intimacy among members than commonly occurs and become disillusioned if this is not the case. Others may assume that more energetic members will keep the community going and that their own active participation will not really be needed. Member participation, however, is critical to the effective functioning of cohousing developments. Respondents strongly advised that potential cohousers consider their desire and ability to take this responsibility seriously, "Be very open to the idea of working actively in the community…That's what I would tell them 'cause a community needs you." (Paige), and "I would just say that they understand the work commitment.. They have to work. And they have to work and play well with others for the best experience." (Dixie). Ava concurred, "You're gonna get out as much as you put in."

Residents agreed that those interested in senor cohousing should ensure that their expectations are realistic. Realize that while there is a strong probability that you will find good friends within the community, it is also likely that there will be others who you will find bothersome. Understand that conflicts will arise, there will be struggles, at times, and that resolution takes hard work by good-hearted people. When speaking of realistic expectations, one resident advised "Make sure that you're seeing things that are real. Well, I had an idea that the people here were super people. In a sense that they wouldn't react the way most people do. They were too good to be true... So, look at things as they really are. They're people and you have to see them that way. They have their faults like everybody else...They get sick. They get well. And they get mad, yeah, they get down. And think that if I were to talk to somebody who has never experienced this, never known, the first thing I would say is make sure you're seeing things that are real." (Patty). Another person recommended, "I'd say be absolutely realistic about people. Don't think you're gonna move in here and we're all gonna love each other and we're all gonna do things together and gather down here every night and watch TV and play games... you have to be really, really be realistic about things. And be responsible for yourself and not depend upon everybody here to make you a happy person." (Tammy, SS).

Be willing to grow. As discussed earlier, there are multiple opportunities for personal growth when living in senior cohousing (Fig. 8.2). However, not everyone is interested in or open to change within themselves or in their way of life. And so, cohousers recommend, "Be sure that you're willing and open to growing, because the opportunities are there, and if you are not willing to grow, personally as well as community, an intentional community may be more challenging than you can handle, because...it's an incredible opportunity, but if you're the kind of person that just is, "I like the way my life is right now, I like living in my own place. I like having my own rules, I like doing it my own way, and I don't have to listen to anybody else," this is not the kind of community for you. It isn't, and that's okay. That's perfectly okay. There's all kinds of communities out there." (Ava). Perhaps one of those most challenging and rewarding areas of growth has to do with letting go of ego. The opportunity to let go of ego usually comes when one is faced with varying options, such as: to push a strong personal preference or go along with the will of the majority; to dominate the workings of a committee or let others take charge; to hold onto power or relinquish control; to persist in one's opinions or consider that another's perspective may have equal value. One woman (Deb) gave an example of her own thinking about ego before she moved into cohousing, "...the concept of trying to prove yourself, or your ego. I mean, you have to let go of that, you know, and I thought about that long and hard, when I moved here, and I thought, when you move there, you don't have to prove anything, you know. I've had a successful life. I've proved myself enough, and I'll let go of that. I don't have to be proving anything to anyone or have my ego up here...we need to keep our ego outside the door when we come into meetings, and that's a hard step. That's a big step."

Fig. 8.2 Participating in discussions with others

8.3 Be Courageous

Making the decision to move into a senior cohousing community may not be an easy one for many. Some who are drawn to living in a SCC may find that they are also tentative about taking this step (Fig. 8.3). After all, for most, cohousing means leaving behind the known, the comfortable, the way things have always been done. As a result, some may hesitate and think, "Perhaps I'll make the move later, when I'm older". However, senior cohousers advise that those who are attracted to life in an SCC make the move sooner rather than later. The primary reason given is that having others around to share your life with is much better than being isolated or lonely and that cohousing is rewarding and fun. Therefore, many suggest that it makes sense to relocate to cohousing when one is relatively young, healthy and able to take advantage of all community living has to offer.

Another practical reason has to do with reciprocity. As described in previous chapters, senior cohousers value the support they receive from and are able to give to others. However, those with existing needs for support should not move in with the expectation that others will take care of them. As Noreen explained, "…if you wait until you need it [support], until you need to move, you've waited too long because you need time to build the social capital". So, senior cohousers advise, doing research, visiting a variety of SCCs, getting to the know the residents at your preferred community, engaging in self-reflection and if after all of this, you feel drawn to life in a particular community, go ahead, and make the move. Ava explains,

Fig. 8.3 Welcoming guests and new residents

"It takes a lot of courage to be happy. It takes a lot of courage to be happy, and to make your life happy, because life is problems, and it takes a lot of courage to get what you want and it's scary. Jenny agreed, "...don't hesitate, have courage. And if you don't have a lot of courage, have a sense of adventure."

In sum, if you are seriously considering a move to a senior cohousing, keep in mind that it is important to visit several communities and experience life within a community by having an extended stay, if possible. Additionally, consider you needs, your disposition and what you are willing to offer. Finally, and importantly, be prepared for growth and a new adventure.

Chapter 9
Living and Learning—The Senior Cohousing Experience

> The heart of the communities is the relationships. (Kim)

> You know, it takes all kinds to make a heaven. We're all very independent and yet we respect each other's independence. We rely on our commitment to each other. (Earl)

After visiting twelve senior cohousing communities, the major question we had as authors were—What did we learn? While there were a number of important findings, which evolved into the previous chapters of this book, there were three overarching themes that came through. First, people who move into Senior Cohousing desire connection beyond what is found in typical neighborhoods and in many other types of senior housing communities. Secondly, being part of a senior cohousing community fosters personal growth and development for those who have open hearts and minds. Finally, there are some important issues for the SCC movement to consider as older adults move forward in developing ever more cohousing communities.

9.1 The Desire for Connection

As the quote by Kim at the beginning of this chapter exemplifies, social connection is the dominating theme within cohousing communities. In every chapter of this book, relationships are discussed in some way. Senior cohousing enables individuals to more easily stay connected to others and avoid the isolation that may arise from living in other types senior housing where the development of community is not a central principle (Fig. 9.1). There is reciprocity in the relationships formed among community members. Not only do members receive emotional and practical assistance when needed, but, importantly, they are also able to provide meaningful help to others. The significance of both receiving and providing aid was a consistent theme in our conversations. Multiple participants noted how these interactions are reminiscent of neighborhoods that were common in their youths but which are now

S. Cummings and N. P. Kropf, *Senior Cohousing*, SpringerBriefs in Aging,
https://doi.org/10.1007/978-3-030-25362-2_9

Fig. 9.1 Connection through shared interests and activities

vanishing due to an increasing mobile and fast-paced society. Marie summed it up this way, "how do we restore a sense of neighborliness and community? … a lot of what is being lived out here is taking care of a neighbor or running to the store or taking someone to the airport. Those are the kinds of things that neighbors used to do."

To be clear, however, the desire for connection does not mean that the residents in senior cohousing want to do everything together! The majority of those interviewed plainly stated that they needed time to be alone, to have some privacy, and to have contacts away from the SCC. Just like other types of relationships, living in a cohousing community often involves spending time together, time alone, and time with others. Those who seemed the most satisfied with the decision of relocating to a SCC had negotiated this balance well. Many participants spoke enthusiastically about activities in which they were involved outside of the SCC. From volunteering, running for office, engaging in artistic pursuits, and working part-time to kayaking, biking, hiking and enjoying cultural events, participants took advantage of opportunities in the larger community. This connection to and engagement with the surrounding area was important for many residents. Many were clear that the desire to move to the particular SCC where they lived included a desire to be part of the larger community. We visited senior cohousing communities in rural areas, smaller cities and towns, and larger urban areas. Although there was variation in locales, the residents discussed how the particular setting in which their SCC was located added to their quality of life and satisfaction with their community.

9.2 Growth and Development

Another theme that ran through many conversations focused on the personal growth and development experienced as a result of living in a SCC. The vast majority of residents agreed that they had grown since coming to live in their cohousing community. They reported that if you really want to make cohousing work for yourself and the community, then you need embrace both the benefits and then challenges as described in Chap. 6. The experience of facing these challenges is how one grows. One person reported that living in cohousing is a "a continuous journey" of growth (Mae). Stephanie agreed "Well, I think community has a lot of benefits, especially as you get older… You give and you take, and so forth. And there's a lot of satisfaction in that because there's a lot of self-growth in it. And so you begin to realize that you become a better person." The growth that residents spoke of took many forms but most centered on developing greater interpersonal skills and cultivating patience, tolerance and the ability to let go (Fig. 9.2).

In Chap. 2, we briefly described the theory of gerotranscendence in which our study was rooted and posed the question of whether living in this type of community helps foster older adults' continued growth. According to Tornstam (2005, 2011), growth in later life occurs on Cosmic, Self and Social/Personal levels. When speaking of their experiences within their SCCs, participants' stories, for the most part, did not speak of change at the Cosmic level. Increased feelings of oneness with the universe, connection to both past and future generations and the ability to see the universal in individual objects, for example, were not themes reflected in their conversations.

Fig. 9.2 Sharing expertise with others

One aspect of the Cosmic arena that was mentioned, however, had to do with death. While no participant specifically mentioned a decrease fear of death, many residents did discuss their increased consideration of aging-in-place, the importance to being able talk about death with other residents, and even the opportunity to celebrate the coming death of a person with a terminal illness. As stated previously, Durrett (2009) claims that potential SCC members' participation in Study Group I activities, which heavily focus on the reality of growing older and aging-in-place, is crucial to the development of community in senior cohousing. Study Group 1 is part of the comprehensive model developed by Henry Neilsen for the creation of senior cohousing communities. Members of some of the communities that we visited had participated in Study Group 1 as part of the communities' early developmental processes. This may have primed some communities to have difficult conversations about aging and death. However, for other communities these conversations seem to occur naturally as individuals experienced their own and others' aging process and witnessed the deaths of community members they held dear. While such experiences transpire for all older adults, senior cohousing seems to present the opportunity for many to process these realities with others whom they trust and to grow in their own understanding and acceptance of the aging process and mortality.

On the level of Self, the major theme discussed was related to a decrease in self-centeredness. Multiple, although not all, residents noted their initial growing realization and then intentional struggle to accept the validity of the perspectives and opinions of others with whom they did not agree. While this struggle was often a frustrating one, residents who discussed this aspect, noted their deliberate development of new strategies to deal with "difficult" members—whether this be practicing patience, disengaging from problematic conversations to consider another's point of view or engaging in research to further understand another's perspective. Some described this growth in terms of letting go of ego and of the need to be in charge. Change in this aspect of self can be one of the hardest we as humans face. Members' open discussion, in some SCCs, of the importance of "leaving their ego at the door" for the good of the community seemed to reinforce the significance of growth in this area.

Most of the change discussed by senior cohousing residents can be found in themes related to the Social/Personal level. When reflecting on change, the majority of residents who responded noted growth in compassion, respect and patience (see Fig. 9.3). Residents reported that development in these areas was necessary to effectively deal with other members. They recognized that being short-tempered or sharp tongued was detrimental to community cohesiveness. Several stated that previously they had not fully recognized how hurtful such behavior was but when living and working with others day in day out, the negative impact of this type of conduct became much more evident. Likewise, deliberate efforts to grow in the areas of tolerance and non-judgmentalism were frequently mentioned by residents. They attributed these efforts to their desire to maintain positive interactions among community members. Respondents noted that, while in the past they could simply ignore or avoid those they didn't like, this was not possible when living in community. Several stated that although they did not particularly like everyone living in their SCC,

Fig. 9.3 Promoting peace and harmony

they did work to love each member and would unhesitatingly provide assistance to each one as needed.

Another theme in the Social/Personal level that was frequently mentioned had to do with decreased attachment to material possessions. A desire to down-size and to have a smaller ecological foot was one of the things that drew many residents to senior cohousing. People mentioned the wisdom of sharing equipment such as lawn mowers, washers and dryers, and spoke with pride about their communal gardens, recycling efforts and energy efficient buildings. While interest in such activities preceded residence in a SCC for many, living in cohousing often enabled members to more readily divest themselves of unneeded possession and more easily engage in environmentally friendly practices.

The stories of those we interviewed suggests that for those who are open to and interested in growth, life in senior cohousing provides many opportunities for increased awareness about one's own behavior, feelings and reactions and multiple ongoing occasions to practice and develop new ways of being. Not all SCC members, however, may be interested in personal growth or change. Community life, therefore, may be more difficult and less rewarding for them as individuals and less beneficial and productive for the development of the group. For this reason, residents we interviewed advised that those interested in senior cohousing seriously consider their adaptive capacity and openness to self-change prior to making a decision. Yet, for older adults who are willing to engage in the challenge and adventure of continued individual and interpersonal growth, senior cohousing may be a wonderful match.

9.3 Thoughts for the Future

In its current form, there are some aspects of senior cohousing life that make this option less accessible to a wide range of older adults. One major issue is the current homogeneity of the residents. Although residents come from many different areas of the country and have a wide variety of religious beliefs, work histories, personalities and backgrounds, the majority do tend to be white, well educated, and affluent. This situation is partially related to the fact SCCs are designed, developed, managed and operated by the older adults themselves and these individuals tend to be those who possess the educational background and financial resources to undertake such as a complex venture. In addition, senior cohousing is still a relatively new phenomenon in the United States and only a limited number presently exist. The individuals who have built and live in these communities are the pioneers of the senior cohousing movement. They are experimenting with features and approaches and are laying the groundwork for the possible creation of a more inclusive environments in the future.

Diversity. Of the residents that we spoke with during the visits, few were from diverse racial and ethnic backgrounds. Several people commented on their desire for greater diversity within the resident population, yet this was not something presently achieved in most communities. Likewise, many of the SCCs we visited were in neighborhoods that are fairly homogeneous. People in these cohousing communities commented that their social networks manly consisted of individuals who looked very much like them. A couple communities, however, were located in middle of large urban areas where there was a great deal of racial, ethnic and socioeconomic diversity.

There were a few notable areas where diversity was currently a part of senior cohousing communities. One was religious and spiritual differences as people noted that they were Christian, Jewish, Buddhist and atheists. For example, one of the authors participated in a vesper service that was a regular part of the weekly calendar. On this day, the reading was from a Christian passage but the service and discussion also included Jewish perspectives. The integration of multiple religious perspectives was viewed as adding richness and texture to the experience. Additionally, there were same sex couples in a few of the SCCs we visited. Most telling, this issue was never highlighted as something unusual. It was only discovered in passing – such as in recounting a story or in giving a tour of where residents lived. In this way, the inclusion of same sex community members seemed very open and accepting.

Costs. Developing senior cohousing is not cheap! There is the need to find and purchase land, hire developers, contractors and architects, pay for all required building materials, permits, etc., and furnish common areas. When these initial costs must be covered by a small group of potential members, many find participation beyond their reach. Several of the communities are located in upscale neighborhoods where land is more expensive. Others possess higher-end materials and more extensive amenities and common areas. These features drive up initial development expenses and the sale prices of individual units and homes. Some groups, however, have found ways to substantially restrict beginning costs. In one SCC, initial expenses were undertaken by a senior housing developer who was an older adult himself and who wished to live in the completed cohousing community. Some communities have received grants to offset development costs while others have worked with non-profit agencies. The founder of one community who had experience as a developer and a friend in real estate oversaw most of the initial work completed in her community. And, a few SCCs that we visited had purposefully chosen to construct smaller communities located in less expensive areas of town. Such approaches made these SCCs more affordable to middle class individuals. Still, buying into a typical senior cohousing community may be beyond the means of many older adults with restricted incomes and limited financial resources.

To address this reality, some of the SCCs have established ways to help lower income older adults afford to live in their communities. One of the communities visited had below market value homes that allowed them to receive a tax break from the local government. These were colloquially called "affordables." The difference in home price between affordables and market value was significant. While individuals who had fewer resources were given the opportunity to move in, there were also some unintended consequences as described by Tammy "There is some resentment on the part of some of the people in affordables …. As the people who live in market rate for the most part are well-to-do people. This is [city] and to own one of our units it's close to a million dollars as opposed to the affordables that are $170,000 and will never be more than that."

Another way that other communities have made living there more affordable is to have a few rental, in addition to the privately owned, properties and to allow individuals to rent out part of their units. On a tour of one community, one of the authors visited several homes that had a floor rented out. A single woman who was living there moved from across the country to be close to her daughter. She was over-the-moon happy that she had a one-bedroom apartment and was able to be part of such a well-off environment that she would not be able afford any other way.

A second community had apartments in the common house that were offered at a reduced price. These were comfortable, self-contained units that included a living area, bedroom and small kitchenette. Jill discussed how she came to live in this apartment and what it meant to her. She described how she was caring for her father, and within a short time she lost him, and her only child also passed away. As a result of the magnitude of these losses, she researched cohousing but discovered that she didn't have the resources to buy a property in a senior cohousing community. "I went and visited them [other cohousing communities]. It was $220,000 to buy a one-bedroom house. I don't have that money, so I was automatically taken out of the of senior cohousing category when I started looking at the pricing. Then I found this place. It was like it came down from somewhere, I don't know. I called, and I found it was mixed income."

Lastly, some non-profit agencies, such as Sarah's Circle featured in this book, are building senior rental properties that are intentionally designed to foster community. With features such as large common areas, participation of residents on the board of directors, resident managers who are community members and encouragement of residents to actively engage in community life, these groups are on the cutting edge of extending many of the benefits of senior cohousing to low income older adults.

9.4 Concluding Thoughts

As we end this book, both authors are left with a rich set of experiences as well as a picture of the lives of senior cohousing residents. Before visiting the communities, meeting the residents, and participating in activities, we had a theoretical understanding of this type of living situation. That is, we assumed that it would be a positive experience since we know that older adults both value being in their own homes and thrive on social engagement. What we didn't expect was the level of resilience and vitality that we encountered! Although we have tried to capture that spirit in these chapters, we want to leave with a few quintessential stories of our time with these amazing individuals.

- One of the authors was supposed to interview Norma (age 68) as she had initially agreed to participate in a conversation. The day of the visit, however, was perfect weather—blue sky, low humidity. Upon arrival, a voice from the roof called out "I'm up here." You see, she was repairing her roof and wanted to make the most of the day. Although she came down the ladder to apologize, it was clear that her priority was to finish the job with her male friend who had come over to help out.
- In a focus group interview one of the participants, Dorothy, who was 90, commented that, at times, she felt bad because she could no longer activity participate on teams dedicated to the continuing upkeep of the cohousing community. The other participants immediately protested and argued that she greatly contributed to community life by being a role model for them of positive aging. Another member of the group jokingly commented that whenever he went out in town with Dorothy, he felt that she was his "arm-candy" since everyone in town knew and liked her and always stopped them to say hello and talk for a while.
- Kim (Age 79) joined the group interview a bit late. She had been out in one of the three organic gardens on the property and was finishing up with weeding and tilling. She also had to tend to one of the chicken coops on the property. Although she was pretty dirty (and a little smelly), she jumped into the discussion as soon as she arrived.
- One of the authors had the opportunity to share a meal with a small group of women in their 90's who all had lost their husbands. The women in this "widows' group" came together every Friday night in the common house kitchen to share dinner with one another made by a cook they had hired. In the middle of a lively conversation, one of the women who was speaking suddenly stopped and said, "I can't remember what I was going to say". The others immediately started laughing and claimed that forgetting something in mid-sentence happened to all of them. After a brief pause someone started speaking and the conversation took off once again.

- Stephanie (Age 89) was the host for one of the author's visits. After the group interview, she was going back to her house and was going to meet the author at the vesper service. She never showed up.... At dinner afterwards, she stated that a friend unexpectedly stopped by and she opened a bottle of wine instead of going to the service. "I've been to church a lot in my life.... Seemed more fun to share some wine today."

Of course, after conducting all of our interviews we realize that living in a communal setting does has some challenges and creates some tension. Overall, however, both authors experienced a great deal more positive perceptions from the residents than problematic outcomes.

We began this book with two purposes. One was to explore the experience of living in a senior cohousing community in order to add to the gerontology literature. As academics, we hope to have a broader perspective of later life living options to share with our students who are the next generation of professionals who will practice with older adults. The inclusion of senior cohousing expands the understanding of later life choices and how activity and engagement benefit older individuals.

A second, and more personal goal, was to explore these communities as aging adults ourselves. There are many reasons that a cohousing arrangement is an attractive possibility. More than anything, we were both overwhelmed with the generosity and openness that we experienced in our travels. While neither of us put down a down payment at this point, both of us are seriously considering this type of living situation as an option in our own futures.

References

Durrett, C. (2009). *Senior cohousing handbook: A community approach to independent living* (2nd ed.). Gabriola Island, BC, Canada: New Society Publishers.

Tornstam, L. (2005). *Gerotranscendence: A development theory of positive aging*. New York: Springer Publishing.

Tornstam, L. (2011). Maturing into gerotranscendence. *The Journal of Transpersonal Psychology, 43*(2), 166–180.

Appendix A: Questions for Members of Senior Cohousing Communities

1. What drew you to the idea of participating in an Senior Cohousing?
   a. What were your fears/concerns, if any, about joining a SCC?
   b. How did you deal with these issues?
2. What was your concept of community before you became involved in your SCC?
   a. Has this changed over time and, if so, how?
   b. What were your expectations of community when you moved in?
3. How do you balance involvement in community with needs/desires for "alone time?
   a. Has this been challenging?
   b. Did achieving this balance for you as an individual or for the community take time/discussion to evolve?
4. Aging-in-place—what has been your own experience?
   a. What has been the community's experience?
   b. How has this changed your view of yourself and your relationship to others?
5. Has the *type* of activities, your *level* of engagement in activities, or *how* you engage in activities changed at all since you began living in your SCC? If so, How? What difference has this made to your life?
6. Are the majority of your activities within the SCC or in the surrounding community? How would you describe the relationship between your SCC and the surrounding community?
7. What are the benefits, challenges, and surprises you've encountered living in your SCC?
8. What are your desires for yourself and for this community for the future?
9. If you were giving advice to someone who was considering a move to an SCC like yours—what would you say?

S. Cummings and N. P. Kropf, *Senior Cohousing*, SpringerBriefs in Aging,
https://doi.org/10.1007/978-3-030-25362-2

10. What did you learn about yourself and how have you grown, if any, through your involvement in this intentional community?
11. Anything I didn't ask that you feel I should have or anything you feel that I should know.

*For those who were involved in planning and developing the SCC?

Did you participate in Henry Nielsen or the Charles Durrett/Jean Nillson study groups to develop and design your community? If so, what was helpful/important about these planning sessions?

a. If not, how did you develop and design your community?
b. What did you learn about yourself through this process?
c. Did your understanding of aging and your future change through this process and, if so, how?

Printed by Printforce, the Netherlands